Contents

CW01501407

Landlords' consents

A practical guide

Daniel Kidd

and

Gabrielle Higgins

Acknowledgements

Crown copyright material is reproduced with the permission of the Controller of HMSO and the Queen's Printer for Scotland.

Please note: References to the masculine include, where appropriate, the feminine.

Published by the Royal Institution of Chartered Surveyors (RICS)

Surveyor Court

Westwood Business Park

Coventry CV4 8JE

UK

www.ricsbooks.com

ISBN 978 1 84219 448 5

Typeset in Great Britain by Columns Design Ltd, Reading, Berks
Printed by Page Bros, Norwich

Preface

Surveyors are often asked to make, or advise on, applications for landlords' consent to assign, sublet, alter or change the use of leasehold premises. Although the principles that apply are complex and often arcane, there seems to be little practical guidance available.

We hope that this book will help to resolve that problem. Its aim is to guide surveyors dealing with properties in England and Wales: to help surveyors to make the right decisions when acting alone for their clients and to indicate when matters may be more complicated and legal advice may be needed.

We could not have produced this book without significant help from a number of sources.

First and foremost, thanks go to Jenny and Mark, who have helped and supported us throughout, in addition to making many useful suggestions on the manuscript. We are also very grateful to several of Daniel's colleagues at Denton Wilde Sapte: Julian Cridge and Katharine Fenn for their encouragement, and Robert Wyatt for reading and commenting on chapter 9. Thirdly, we owe a debt of gratitude to all those at RICS Books, and in particular Sophie Brooks, for all their help and tolerance.

In terms of written material, Woodfall's *Landlord and tenant* and the first edition of what was then Letitia Crabb's *Leases: covenants and consents* have both been invaluable.

Despite the help we received, all the errors that remain are our own. We have aimed to state the law as at 1 December 2008.

Daniel Kidd and Gabrielle Higgins

December 2008

List of Acts, Statutory Instruments and abbreviations

The following Acts and Statutory Instruments are referred to in this publication. Where an Act is mentioned frequently, it is referred to by the abbreviation in brackets that follows.

Access to Neighbouring Land Act 1992

Agricultural Holdings Act 1986

Civil Partnership Act 2004

Commonhold and Leasehold Reform Act 2002 (**'CLRA 2002'**)

Companies Act 1985

Companies Act 2006

Communications Act 2003

Crime and Disorder Act 1998

Criminal Justice Act 1991

Criminal Justice and Public Order Act 1994

Criminal Law Act 1977

Crown Proceedings Act 1947

Disability Discrimination Act 1995

Equality Act 2006

Factories Act 1961

Fire Precautions Act 1971

Housing Act 1980 (**'HA 1980'**)

Housing Act 1985 (**'HA 1985'**)

Housing Act 1988 (**'HA 1988'**)

Housing Act 1996 (**'HA 1996'**)

Human Rights Act 1998

Immigration and Asylum Act 1999

Insolvency Act 1986

Land Registration Act 1925

Land Registration Act 2002

Landlord and Tenant Act 1927 (**'LTA 1927'**)

Landlord and Tenant Act 1954 (**'LTA 1954'**)

Landlord and Tenant Act 1987 (**'LTA 1987'**)

Landlord and Tenant Act 1988 (**'LTA 1988'**)

Landlord and Tenant (Covenants) Act 1995 (**'LT(C)A 1995'**)

Law of Distress Amendment Act 1908

Law of Property Act 1925 (**'LPA 1925'**)

Law of Property (Miscellaneous Provisions) Act 1989

Leasehold Reform Act 1967

Leasehold Reform, Housing and Urban Development Act 1993

Limitation Act 1980

Local Government and Housing Act 1989

Matrimonial and Family Proceedings Act 1984

Matrimonial Causes Act 1973

Offices, Shops and Railway Premises Act 1963

Party Wall etc Act 1996

Protection From Eviction Act 1977

Race Relations Act 1976

Rent Act 1977 **('RA 1977')**

Sex Discrimination Act 1975

Sexual Offences Act 1956

Shops Act 1950

Sunday Trading Act 1994

Telecommunications Act 1984

Town and Country Planning Act 1990

Tribunal, Courts and Enforcement Act 2007

Disability Discrimination (Employment Field) (Leasehold Premises) Regulations 2004

Disability Discrimination (Providers of Services) (Adjustment of Premises) Regulations 2001

Equality Act (Sexual Orientation) Regulations 2007

Town and Country Planning (Use Classes) Order 1987

Table of cases

Introduction: How to approach landlords' consents

The key to dealing with landlords' consents is to approach the issues in the right order. This introduction aims to help by providing:

- an overview of the key principles; and
- flowcharts showing the order in which landlords and tenants should tackle them. The flowcharts give references to the relevant sections of the book for further reading.

Key principles

A. Always begin by reading the lease to check for restrictions.

- The usual starting position is that the tenant is free to deal with leasehold property as he wishes, but this starting position is almost always altered in the lease or elsewhere.
- If consent is needed, it will often be the landlord who must consent; but the tenant may also need consent from a superior landlord or management company.
- Although the restrictions in the lease are likely to be the most important, there may also be restrictions in licences or other deeds which affect the tenant's plans.

See chapter 1 for further information.

B. It is important to understand the difference between the three main types of restriction:

- 'absolute': a complete prohibition;
- 'qualified': the landlord may give consent but need not act reasonably;
- 'fully qualified': the landlord must not unreasonably withhold consent.

Chapter 1 explains this in more detail.

C.   Even absolute restrictions can sometimes be overcome. In particular, the law often allows tenants to carry out alterations despite an absolute restriction in the lease. Tenants can also sometimes apply to have restrictions modified or discharged completely.

See the section headed 'Overcoming Restrictions' in chapters 2 (on alienation), 3 (on alterations) and 4 (on change of use) for more information.

D.   Qualified restrictions:

- sometimes prevent the landlord from demanding a premium; and
- may be transformed automatically into 'fully qualified' restrictions

See chapters 2, 3 and 4.

E.   Tenants should make a suitable application to begin the process of seeking consent. The application:

- should be in writing, be sent from (or on behalf of) the tenant, and be addressed to the landlord or 'RTM' company;
- should contain sufficient information about the tenant's proposals; and
- should be 'served' in the correct way.

See chapter 5 for further guidance.

F.   Under fully qualified restrictions, there are rules about how landlords must respond. Landlords may have obligations to:

- respond within a certain time;
- reply in writing, giving reasons; or
- pass the application on to others.

Others who receive the application may have similar duties. Failure to comply with the obligations may mean that consent is unreasonably withheld. Landlords may be entitled to ask for more information before responding but should be careful not to ask for too much. See chapter 6 for more details.

G. Under fully qualified restrictions, landlords must have reasonable grounds to refuse consent. They may only impose reasonable conditions when granting consent.

See chapter 7 for a guide to the reasonableness of common grounds and conditions.

H. There may be restrictions on the degree to which landlords may demand costs and on the amounts they can recover. See chapter 8.

I. Landlords should be wary of granting consent inadvertently. See chapter 9.

J. An unreasonable withholding of consent allows the tenant to go ahead anyway. Tenants can also go to court for confirmation that consent has been withheld unreasonably, and may be entitled to compensation. See chapter 10 for further information.

K. Landlords have their own remedies if the tenant breaches the lease. These include claiming compensation, seeking an injunction, or ending the lease. See chapter 10.

Tenants' flowchart

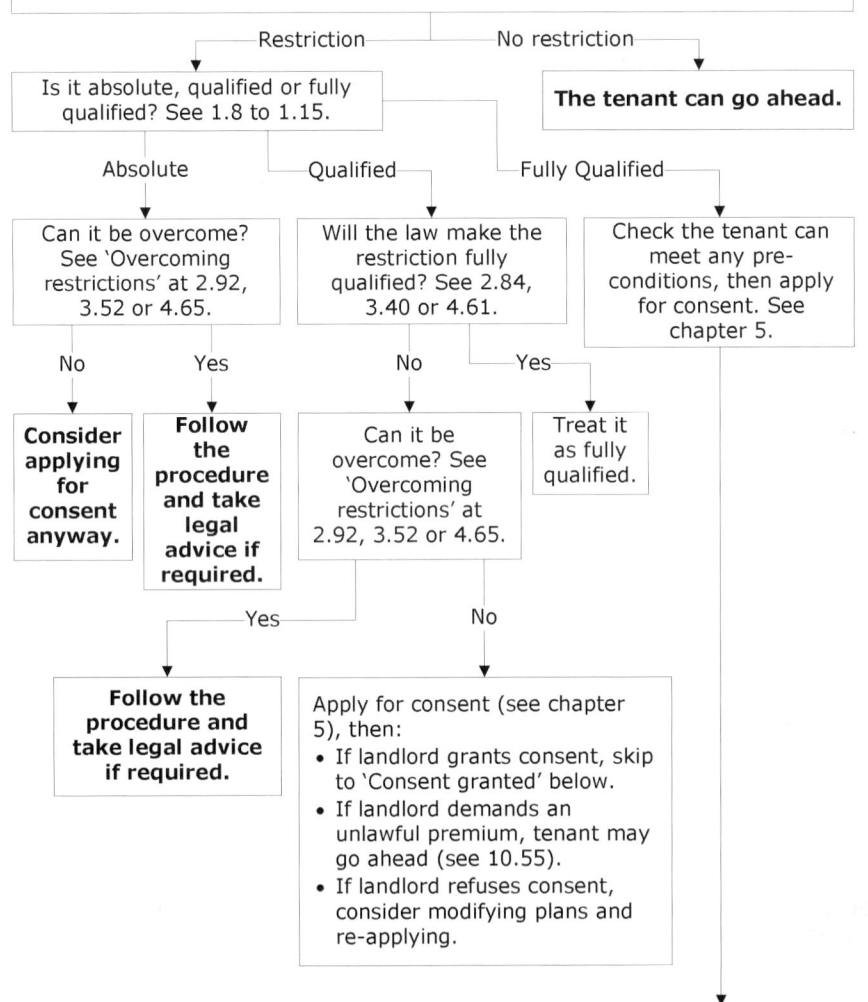

Start by looking at the relevant chapter on:
- alienation (chapter 2);
- alterations (chapter 3); or
- changing use (chapter 4).

Follow the checklists at the end of each chapter to see if there are any express or implied restrictions and whether they cover the tenant's plans.

Restriction ── No restriction

Is it absolute, qualified or fully qualified? See 1.8 to 1.15.

The tenant can go ahead.

Absolute ── Qualified ── Fully Qualified

Can it be overcome? See 'Overcoming restrictions' at 2.92, 3.52 or 4.65.

Will the law make the restriction fully qualified? See 2.84, 3.40 or 4.61.

Check the tenant can meet any pre-conditions, then apply for consent. See chapter 5.

No ── Yes   No ── Yes

Consider applying for consent anyway.

Follow the procedure and take legal advice if required.

Can it be overcome? See 'Overcoming restrictions' at 2.92, 3.52 or 4.65.

Treat it as fully qualified.

Yes ── No

Follow the procedure and take legal advice if required.

Apply for consent (see chapter 5), then:
- If landlord grants consent, skip to 'Consent granted' below.
- If landlord demands an unlawful premium, tenant may go ahead (see 10.55).
- If landlord refuses consent, consider modifying plans and re-applying.

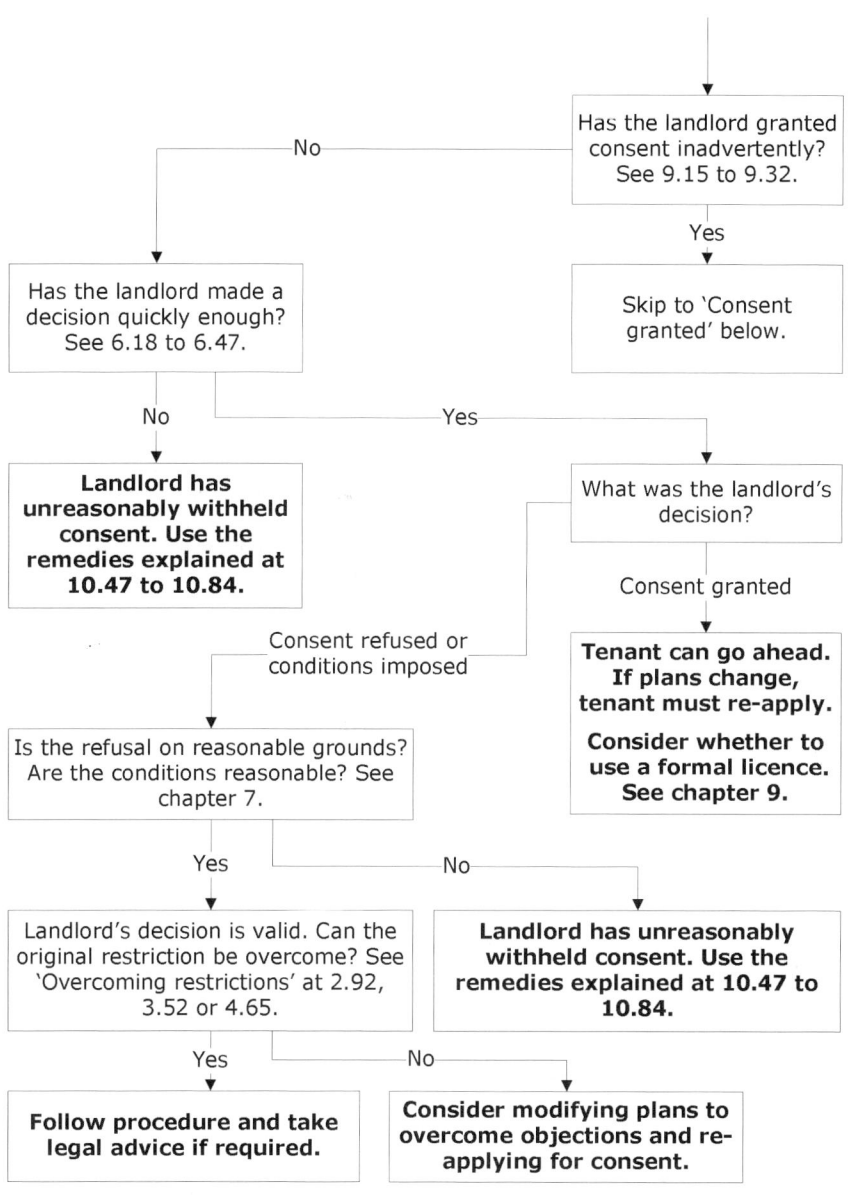

Landlords' flowchart

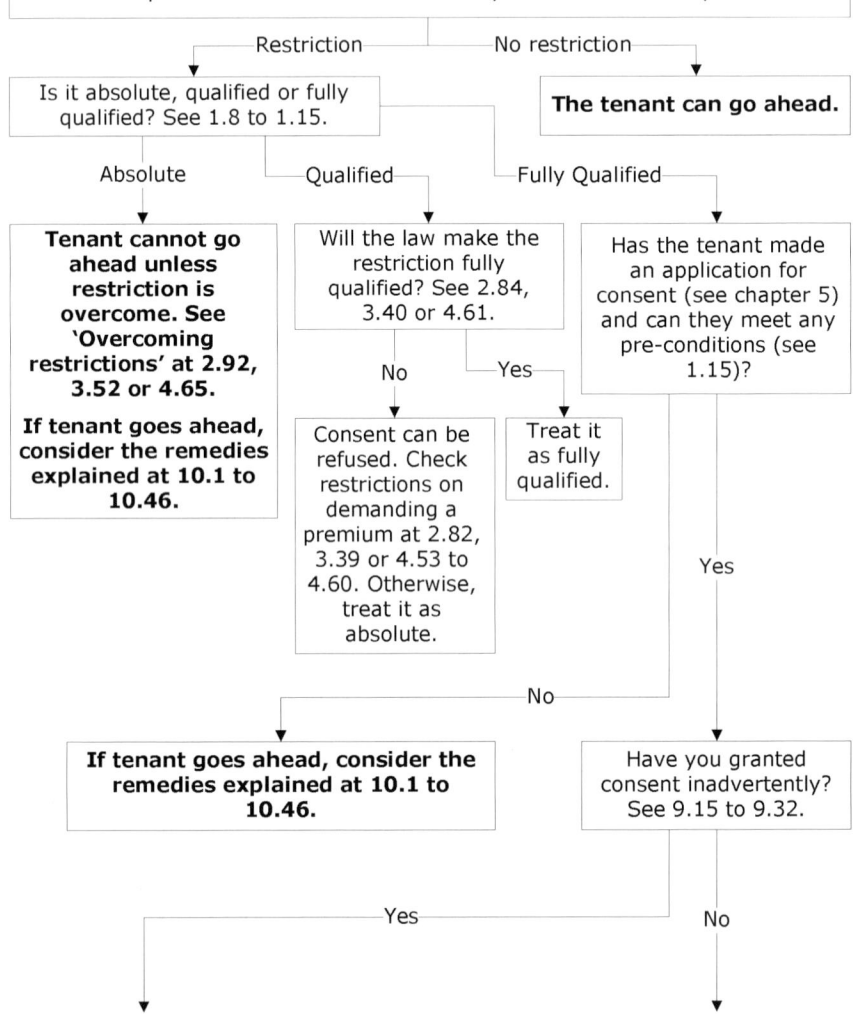

Start by looking at the relevant chapter on:
- alienation (chapter 2);
- alterations (chapter 3); or
- changing use (chapter 4).

Follow the checklists at the end of each chapter to see if there are any express or implied restrictions and whether they cover the tenant's plans.

Restriction ——— No restriction

Is it absolute, qualified or fully qualified? See 1.8 to 1.15.

The tenant can go ahead.

Absolute ——— Qualified ——— Fully Qualified

Tenant cannot go ahead unless restriction is overcome. See 'Overcoming restrictions' at 2.92, 3.52 or 4.65.

If tenant goes ahead, consider the remedies explained at 10.1 to 10.46.

Will the law make the restriction fully qualified? See 2.84, 3.40 or 4.61.

No ——— Yes

Consent can be refused. Check restrictions on demanding a premium at 2.82, 3.39 or 4.53 to 4.60. Otherwise, treat it as absolute.

Treat it as fully qualified.

Has the tenant made an application for consent (see chapter 5) and can they meet any pre-conditions (see 1.15)?

Yes

No

If tenant goes ahead, consider the remedies explained at 10.1 to 10.46.

Have you granted consent inadvertently? See 9.15 to 9.32.

Yes

No

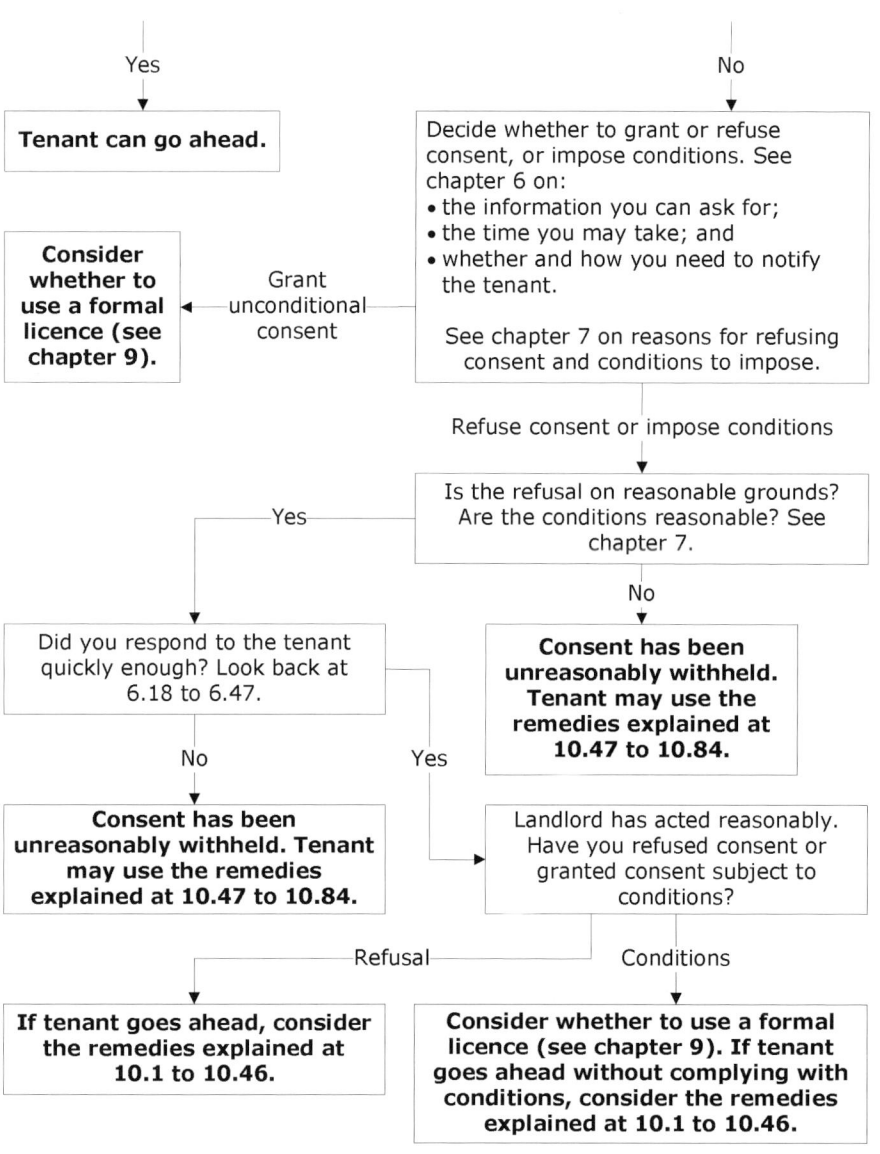

Yes

Tenant can go ahead.

Consider whether to use a formal licence (see chapter 9).

Grant unconditional consent

No

Decide whether to grant or refuse consent, or impose conditions. See chapter 6 on:
• the information you can ask for;
• the time you may take; and
• whether and how you need to notify the tenant.

See chapter 7 on reasons for refusing consent and conditions to impose.

Refuse consent or impose conditions

Is the refusal on reasonable grounds? Are the conditions reasonable? See chapter 7.

Yes

No

Did you respond to the tenant quickly enough? Look back at 6.18 to 6.47.

Consent has been unreasonably withheld. Tenant may use the remedies explained at 10.47 to 10.84.

No

Yes

Consent has been unreasonably withheld. Tenant may use the remedies explained at 10.47 to 10.84.

Landlord has acted reasonably. Have you refused consent or granted consent subject to conditions?

Refusal

Conditions

If tenant goes ahead, consider the remedies explained at 10.1 to 10.46.

Consider whether to use a formal licence (see chapter 9). If tenant goes ahead without complying with conditions, consider the remedies explained at 10.1 to 10.46.

1

Identifying the need for consent

This chapter explains:

- the importance of reading the lease;
- whether the tenant needs consent;
- the LTA 1988 and when it applies;
- whether the lease can specify when consent can be refused;
- whose consent is needed;
- which covenants affect the tenant's plans; and
- how to analyse covenants.

The importance of reading the lease

1.1   There may be many restrictions on a tenant's freedom of action in relation to his premises. For example, there may be planning restrictions or, if the lease is mortgaged, restrictions imposed by the lender. This book is concerned with only one set of restrictions: those between the landlord and the tenant.

1.2   The usual starting position is that the tenant is free to deal with the property as he wishes: he can assign, sublet, make alterations, change use and so on without needing to ask his landlord. But this starting position is almost always altered in the lease or elsewhere.

1.3   Most leases have to be created by deed and so there will be a written document setting out the terms of the agreement between the landlord and the tenant. The main exception is leases for three years or less at a full rack rent which start immediately.[1] But even where the lease does not have to be created by deed, there is usually a written document.

1.4 Almost all modern leases contain terms about assigning, subletting, making alterations, changing use and so on. It is these terms which largely dictate what the parties can and cannot do. A landlord is almost never able to prevent the tenant from doing something against the landlord's own interests, unless the tenant has agreed not to do it in the lease (or a restriction is imposed by some other document or by law).

These terms are often in standard form, and so it can be tempting to assume that any restrictions are the same as those in another lease, or to rely on someone else's summary of what the lease says. However, leases are full of small variations, and even a minor change can affect the landlord's or the tenant's position. It is therefore vital always to check what the lease actually says before taking any action. This applies both to tenants and landlords: tenants must check before making a change or applying for consent to do so, and landlords must check before responding to an application or objecting to something the tenant has done.

1.5 Chapters 2, 3 and 4 explain in more detail what a tenant can and cannot do in relation to assignment and subletting, making alterations, and changing use, both if there are no restrictions, and by reference to common lease terms.

Whether the tenant needs consent

1.6 As explained above, the tenant's freedom of action is usually restricted by express terms in the lease. Restrictions are usually found in the tenant's covenants. They fall into a number of different categories, as explained below.

1.7 It is important to understand the type of restriction that applies in each case. This is because the restriction dictates whether:

- the proposed act is completely prohibited; or
- the tenant may do the proposed act but he must get approval first; and if so whether the landlord must act reasonably when considering the application for consent.

Absolute, qualified and fully qualified covenants

Absolute covenants

1.8 An absolute covenant completely prohibits the act it describes. The tenant may ask the landlord to consent, but this will not be governed by the lease and the landlord may wilfully ignore or refuse any application that is made. The landlord may often demand a premium in return for giving consent.

Example

'Not to assign part only of the demised premises'

This clause absolutely prohibits assignments of part of the premises. However, because it does not refer to other types of alienation, the tenant is free to do anything else such as assign the whole or sublet (provided it is not restricted by another clause or impliedly restricted) without breaching the lease or needing the landlord's consent.

If the tenant applies for the landlord's consent to assign part, the landlord could refuse for any reason, and would not even need to respond.

1.9 An absolute covenant can only be overcome in one of two ways. These are:

- *Where the landlord freely agrees to grant a variation of the lease.* The landlord cannot be forced to do this, and can demand anything it wishes as a condition of giving its agreement. In reality, the landlord's willingness to vary the lease will depend on various factors, such as how important the variation is to the tenant, the market for the premises, whether the variation might benefit the landlord (for example, if the tenant wants to assign the lease to a blue chip company), and whether the tenant is willing to pay any money to the landlord.

- *Where the law specifically provides a way to overcome the restriction.* Exceptions like this are usually very limited and tenants must be careful to comply with any terms or procedures that are laid down. This is explained further at 2.92 to 2.100, 3.52 to 3.105 and 4.65 to 4.73.

Qualified covenants

1.10 A qualified covenant requires the tenant to obtain someone's consent before he can lawfully do the restricted act.

Examples

(1) 'Not to use the demised premises otherwise than as a sandwich bar without the landlord's prior written consent'

Unless there are other restrictions, this clause permits the tenant to use the premises as, for example, an estate agency, provided he gets the landlord's written consent in advance. Every use other than as a sandwich bar requires consent.

(2) 'Not to use the demised premises for a sale by auction without the landlord's prior written consent'

This clause permits the tenant to hold an auction provided he gets the landlord's written consent in advance. Again, unless there are other restrictions, it leaves the tenant free to make use of the premises in any other way without the need for consent.

1.11 The consent required is usually the consent of the tenant's own landlord, but this is not always the case. Sometimes someone else's consent is required as well or instead – for example, a superior landlord or a management company. In this book we will usually assume that it is the landlord's consent which is required. For details on when someone else's consent is required, see 1.53 to 1.54.

1.12 Additional words are often implied, as described in 1.13. Also, the landlord is often prohibited from asking for a premium. See 2.79 to 2.91, 3.35 to 3.51 and 4.50 to 4.64.

Fully qualified covenants

1.13 A qualified covenant might expressly state that the landlord's consent must not be unreasonably withheld. Alternatively, these words might be implied by law, as described at 1.18 to 1.21. We refer to these covenants as 'fully qualified'.

Example

> 'Not without the landlord's prior written consent (such consent not to be unreasonably withheld) to make any structural alterations to the demised premises'

This clause permits structural alterations to the demised premises provided the tenant gets the landlord's consent in advance. The landlord may not withhold consent unreasonably. It leaves the tenant free to make non-structural alterations without breaching the lease or needing the landlord's consent.

1.14 For the importance of the distinction between qualified and fully qualified covenants, see 1.16 to 1.17.

Hybrid covenants

1.15 A hybrid covenant can have features of more than one type of covenant and needs to be analysed carefully. Usually conditions can be identified: if the condition is not satisfied, the covenant is to be regarded as absolute, but once the condition has been satisfied, it is to be regarded as qualified or fully qualified. See also 1.36 to 1.40.

Examples

(1) 'Not without the landlord's prior written consent to use the demised premises otherwise than as a hairdresser (such consent not to be unreasonably withheld in the case of use falling within Class A1 of the schedule to the Town and Country Planning (Use Classes) Order 1987)'

This example permits use as a hairdresser, contains a fully qualified covenant against other A1 use, and contains a qualified covenant against all non-A1 uses. The tenant can apply to change the use to a different A1 use, such as a retail shop, and the landlord will only be entitled to withhold consent if it is reasonable to do so. But if the tenant applies to change the use to a non-A1 use, such as a hot food takeaway, the landlord will be able to refuse consent on any grounds or no ground, and will not even need to reply at all. However, he will not be able to demand a premium unless there are structural alterations involved: see 4.53 to 4.60.

(2) 'Not to sublet the premises except at a full market rent **and** with the landlord's prior written consent, such consent not to be unreasonably withheld'

This example contains two elements. The first is an absolute covenant against subletting the premises at a rent other than a full market rent. The second is a fully qualified, more general, covenant against subletting. Because sublettings at less than market rent are absolutely prohibited by the first element, the second element will be irrelevant if the proposed sub-rent is too low: the tenant cannot sublet anyway. But if the proposed sub-rent is a market rent, the tenant will still not be able to sublet unless he gets his landlord's consent, although that cannot be withheld unreasonably.

(3) 'Not to sublet the premises except at a full market rent **or** with the landlord's prior written consent, such consent not to be unreasonably withheld'

This example is almost identical to the previous example and also contains two elements, but this time the elements are alternatives. The first element gives the tenant complete freedom to sublet at a full market rent, without needing the landlord's consent (unless there are other restrictions elsewhere). The second element will therefore be irrelevant if the proposed sub-rent is a market rent. This is the opposite of the previous example: the tenant can sublet anyway. In addition, if the proposed sub-rent is less than market rent, the tenant will be able to sublet unless the landlord withholds consent reasonably.

(4) 'Not to assign the demised premises without the consent of the landlord, such consent not to be withheld in the case of a respectable and responsible assignee'

This example again contains two elements. The first element is a standard qualified covenant which the law treats as fully qualified, as explained at 2.84. The second element is a proviso that the landlord will not withhold consent if the proposed assignee is respectable and responsible. This means that if the proposed assignee is respectable and responsible, the landlord may not refuse consent even if he could reasonably refuse on other grounds. If the proposed assignee is not respectable and responsible, the landlord may refuse consent if he can do so reasonably, which he is likely to be able to do.[2]

(5) 'Not to assign the demised premises without the consent of the landlord, such consent not to be unreasonably withheld in the case of a respectable and responsible assignee'

This example looks like a hybrid covenant, but in fact it is a simple qualified covenant which the law makes fully qualified, as explained at 2.84. Because the law implies a proviso that consent to assign is not to be unreasonably withheld, consent cannot be withheld unreasonably in any situation, whether the proposed assignee is respectable and responsible or not.[3]

Qualified and fully qualified covenants compared

1.16 The distinction between qualified and fully qualified covenants is of huge importance. Qualified covenants do not impose any obligation on the landlord to act reasonably. A landlord faced with an application under a qualified covenant may refuse consent on any grounds whatsoever, and need not even specify any grounds at all. There is no time limit within which a response should be given and there is no action the tenant can take if he is unhappy with the decision. In most cases, qualified covenants therefore offer little more protection than absolute covenants, because the landlord is not obliged to give consent. The only protection they offer is that the landlord is frequently unable to demand a premium in return for granting consent.

For information on when the landlord can ask for a premium, see 2.79 to 2.91, 3.35 to 3.51 and 4.50 to 4.64.

In limited circumstances, there may be an argument that the landlord must act reasonably, even with a qualified covenant. See 4.61.

1.17 Fully qualified covenants offer a great deal more protection for the tenant. Provided the tenant makes a proper application to the landlord, consent cannot be unreasonably withheld.

Transformation of qualified covenants

1.18 Many qualified covenants are automatically transformed by law. There are two types of transformation:

- the prohibition of a premium; and
- the implication of a proviso that consent shall not be unreasonably withheld.

There is some overlap: if the covenant is fully qualified, it will usually be unreasonable for the landlord to demand a premium, even if a premium is not otherwise prohibited.

1.19 The types of qualified covenant which are transformed and the transformations which are made are summarised in this table. It is a summary only, and reference should be made to the explanations at 2.79 to 2.91, 3.35 to 3.51 and 4.50 to 4.64. It assumes that the covenant is modified by law.[4] See Appendix 2 for full details of when the modifications are made. If they are not, a qualified covenant will not be made fully qualified, and the landlord is often free to demand a premium.

	Qualified covenant against:		
	Assignment and subletting	**Alterations**	**Changes of use**
Is the covenant made fully qualified?	Yes[5]	Yes if the alteration is an improvement, otherwise no[6]	No
If the lease does not mention premiums, can the landlord demand one?	No[7]	No if the alteration is an improvement, otherwise yes[8]	Yes if the change of use involves structural alterations, otherwise no[9]
Can the lease expressly permit premiums?	Yes if carefully drafted (see 2.83 and 2.90)[10]	Yes but the drafting must be careful if the alteration is an improvement (see 3.50)[11]	Yes if the change of use involves structural alterations, otherwise yes if carefully drafted (see 4.54)[12]

	Assignment and subletting	Alterations	Changes of use
Payments not prevented	Legal or other expenses[13]	Damage to value of landlord's premises and legal or other expenses[14]	Damage to value of landlord's premises and legal or other expenses[15]

1.20 Where the law transforms a qualified covenant to a fully qualified covenant, it is impossible to 'contract out'[16] except in limited ways. Therefore an attempt in the lease to avoid the transformation (whether deliberate or not) will usually have no effect. Further, if the transformation of the covenant is more favourable to the tenant than the contractual provisions of the lease, the tenant will be able to rely on the transformation.[17]

> For information about how the transformation can be partly avoided, see 1.33 to 1.52.

1.21 However, only attempts to avoid the transformation to the detriment of the tenant are made invalid. If the landlord and the tenant agree to go beyond the transformation to the benefit of the tenant, the additional terms will be valid.

Example

A tenant covenanted '*not to assign the demised premises without the consent in writing of the landlord such consent will not be withheld in the case of a respectable and responsible person*'.

If the assignee was a respectable and responsible person, the landlord could not withhold consent, even if he had another good reason to do so.[18]

The *Landlord and Tenant Act* 1988

1.22 In the case of alienation covenants, the tenant's position is further improved by the LTA 1988.

1.23 Before the Act came into force, there were many shortcomings in the protection given to tenants by fully qualified covenants.[19] The landlord was under no positive obligation to respond to an application promptly, to give consent unless he had good reason for withholding it, or to say what his reasons for withholding consent were. This made it very difficult for a tenant to assess whether consent had been withheld unreasonably or not. This is still the position in cases where the Act does not apply (for example, for alterations and change of use covenants).

1.24 As explained at 10.51 to 10.54, if consent is withheld unreasonably, the tenant can go ahead with whatever he applied to do without being in breach of covenant. But if he is wrong about whether the landlord's refusal was unreasonable, he will be in breach of covenant. He may be liable to pay damages to the landlord, and his lease may be forfeited. If he wants to assign, the assignee may be unwilling to take a lease which might be forfeited, or to wait while the tenant obtains a court declaration on whether consent has been withheld unreasonably. It can also be difficult to persuade a landlord to respond quickly; even if the landlord would eventually give consent, the assignee may be unwilling to wait very long for a decision.

1.25 In cases where the LTA 1988 applies, the tenant's position is much improved:

- The landlord is obliged to respond, in writing, within a reasonable time.
- The landlord is obliged to give consent, except where it is reasonable not to.
- If the landlord gives consent, he is obliged to specify any conditions (which must of course be reasonable ones) which the consent is subject to.
- If the landlord refuses consent, he is obliged to state his reasons.
- If the landlord fails to comply with any of his obligations, he may have to compensate the tenant for any loss this causes.

See 6.20 to 6.29 for a fuller explanation of the landlord's duties, and 10.67 to 10.79 on compensating the tenant for loss.

1.26 Where the covenant is in a subtenancy, the Act helps the subtenant get any consent which is needed from the superior landlord under a fully qualified covenant, as well as from the landlord. This is explained at 6.9 to 6.17. Where there is an RTM company (explained at 5.23 to 5.25), the Act is modified.[20] The effect of the modifications is unclear, and legal advice should be taken.

When does the LTA 1988 apply?

1.27 The LTA 1988 applies to almost all tenancies, including agreements for a tenancy, even if they were entered into before the Act came into force. It does *not* apply to:

- secure or introductory tenancies;[21] or
- leases granted to contractors in respect of the running of prisons,[22] secure training centres[23] or removal centres.[24]

A secure tenancy is a type of public sector residential tenancy. For more details, see Appendix 1. An introductory tenancy is another type.

1.28 Even in applicable tenancies, the Act does *not* apply to absolute covenants or to qualified covenants which are not subject to a proviso that consent may not be withheld unreasonably.

1.29 Subject to those exceptions, the Act applies to all fully qualified covenants against:

- assigning;
- subletting;
- charging; or
- parting with possession of

the demised premises or any part of them.

It does not apply to other forms of alienation covenant, for example a covenant against making a declaration of trust. Nor does it apply to covenants outside the landlord and tenant situation – for example, where the consent of the mortgagee is required under the terms of a mortgage.[25]

> For further details on common types of alienation covenant, see 2.12 to 2.33,

1.30 The Act applies whether the proviso which says that the landlord's consent cannot be unreasonably withheld is expressly set out in the lease or is implied by law (as described at 1.18 to 1.21).

1.31 It also applies regardless of whether or not there is any other qualification.[26] This means that the Act can apply to covenants against specific subgroups of transaction.

Example

A lease contained covenants:

(a)   not to underlet at less than market rent;

(b)   not to underlet without landlord's consent, not to be unreasonably withheld; and

(c)   'not to underlet the Demised Premises at a rent less than the Basic Rent for the time being payable under these Presents unless the Landlord shall otherwise agree (such agreement not to be unreasonably withheld)'.

The Act plainly did not apply to (a), because that was an absolute covenant.

Equally, it plainly did apply to (b), the general fully qualified covenant, so the landlord could be liable to pay damages if it withheld consent unreasonably to an underletting which was at both market rent and Basic Rent (or more).

The court held that the Act also applied to (c), the more specific fully qualified covenant against underletting at less than the Basic Rent: if the landlord withheld consent unreasonably to an underletting at less than the Basic Rent, it could be liable in damages, provided the underletting was at a market rent.[27]

1.32 Later chapters explain what is 'reasonable', the other duties of the landlord, and what happens if the landlord acts unreasonably. See the flowcharts at the beginning of this book for further guidance.

Whether the lease can specify when consent can be refused

1.33 Where the law makes a qualified covenant fully qualified, the general rule is that an attempt to set out in the lease when consent may be withheld will be invalid. So, for example, specifying that a refusal on certain grounds will be deemed reasonable or specifying that it will be reasonable for the landlord to impose certain conditions will often be of no effect.

Example

An 'old' tenancy contained a fully qualified covenant against assigning, underletting or parting with possession, but provided that if the landlord offered to accept a surrender of the tenancy at the same time as he refused consent, his refusal would not be deemed unreasonable and the tenant would be obliged to surrender the tenancy.

It was held that the proviso was ineffective, and the landlord's refusal to consent was unreasonable. The attempt to specify when consent could be withheld did not succeed.[28]

Most tenancies granted in or after 1996 are 'new' tenancies. Other tenancies are usually called 'old' tenancies. See Appendix 1 for more details.

1.34 However, the lease *can* set out when consent can be withheld in three situations:

- where there is a pre-condition;
- assignments of non-residential new tenancies; and
- change of use and other limited covenants where the law does not transform a qualified covenant into a fully qualified one.

These are discussed in turn below.

21

1.35 This general rule also applies where the lease expressly says that consent is not to be unreasonably withheld, if the law would transform it if it did not contain that provision expressly. For example, a covenant may say *'not to assign without landlord's consent, such consent not to be unreasonably withheld'*. As the law would transform it anyway, the general rule applies and the lease cannot set out what is reasonable.

See Appendix 2 for full details of which covenants are transformed and so are affected by the general rule.

Pre-conditions

1.36 Sometimes statements in a lease which have the same effect as those described at 1.33 will be interpreted as pre-conditions. A pre-condition is a requirement that a certain event or state of affairs should occur before a landlord is required to consider an application for consent. These will usually be valid, despite the general prohibition on setting out in the lease when consent may be withheld. However, in the case of a 'new' tenancy, a pre-condition which attempts to exclude, modify or frustrate the operation of the LT(C)A 1995 will be void. This is explained at 1.46.

1.37 A pre-condition is valid because it does not modify the proviso that consent will not be withheld unreasonably. Instead there is a hybrid covenant (see 1.15). If the pre-condition is not satisfied, then the covenant is an absolute one; the tenant can ask the landlord to consent, but the landlord is under no obligation to deal with his application. If the pre-condition is satisfied, then the covenant becomes a fully qualified one.

Example

An 'old' tenancy contained a fully qualified covenant against assigning. It also contained a provision that if the tenant wanted to assign, he should first make an irrevocable written offer to surrender the lease for no premium. The landlord had 21 days to accept the offer, failing which it would be deemed rejected.

Unless and until an offer to surrender was made, the clause absolutely prohibited assignment. Because the prohibition was absolute, the law did not transform it into a fully qualified covenant. If the offer of surrender was made and accepted, the question of consent to assignment would not arise.

If the offer of surrender was made and refused, then the clause became a fully qualified covenant. The tenant could make his application to assign and the landlord could only refuse if it was reasonable to do so.

The effect was exactly the same as in the previous example. However, because it was worded differently, the clause was held to be valid.[29]

This case involved residential premises. In the case of business premises, an agreement to surrender may well be void by reason of s. 38A of the LTA 1954. An agreement to surrender may also be void by reason of s. 25 of the LT(C)A 1995 in the case of a 'new' tenancy, or, in the case of unregistered leases, for failure to register. Legal advice should be taken.

1.38 Depending on the wording, pre-conditions need not always be satisfied before the application for consent is made. Conditions which can only be satisfied after the application is made, or even after consent is granted, can also be valid.

Example

An 'old' tenancy contained a covenant '*not to underlet the demised premises without the previous consent in writing of the Landlord (which consent shall not be unreasonably withheld) PROVIDED ALWAYS that any permitted Underlease shall be granted subject to like covenants and conditions as are herein contained except as to the rent thereby reserved and the length of the term thereby granted*'. This provision was valid.[30]

In this case there was a pre-condition that the underlease had to be on the same terms as the lease. In other words, the covenant was hybrid: there was an absolute covenant against underletting on terms which were not the same as the lease, and a fully qualified covenant against underletting on terms which were the same.

If the tenant makes an application without making clear whether a pre-condition will be met, the landlord should ask. If the landlord grants consent, he should make clear that he is consenting only on the basis that it will be.

1.39 Sometimes the landlord will be able to decide whether or not to apply the pre-conditions only *after* the tenant has decided that he wants to assign. These can still be valid pre-conditions.

Example

A lease contained a covenant *'not to assign the whole of the demised premises without first having obtained the Landlord's written license which subject to compliance with the following requirements shall not be unreasonably withheld'*. The requirements included provision of a surety for a limited liability company assignee if the landlord so required.

The court held that the clause was not intended to set out a condition that the landlord could impose for the giving of consent to an assignment or to set out a circumstance in which a refusal of consent would be deemed to be reasonable. Instead, the clause was aimed at an earlier stage than the giving of consent. It was laying down conditions that had to be fulfilled before the tenant could apply to the landlord for consent.[31]

The clause was to be interpreted as an absolute covenant against assigning to a limited liability company without provision of a surety, if the landlord required a surety.

1.40 As the previous example shows, it can often be very difficult to tell whether a lease will be interpreted as containing valid pre-conditions, or invalid specifications about reasonableness. In cases of doubt, legal advice should be sought.

Assignment of non-residential 'new' tenancies

Most tenancies granted in or after 1996 are 'new' tenancies. See Appendix 1 for more details.

1.41 Different rules apply to assignment (including parting with possession on assignment) of the whole or part of a 'new' tenancy that is not a residential lease. For this purpose, a residential lease is a lease by which a building or part of a building is let wholly or mainly as a single private residence.[32]

1.42 The landlord and the tenant may agree:

- circumstances in which the landlord may refuse consent to an assignment; or

- conditions subject to which the consent may be granted.[33]

In this book, we refer to these agreements as 'agreements on reasonableness'. An agreement on reasonableness may be contained in the lease itself but need not be. Nor need it be made at the same time as the lease; it may be made at any time before the tenant applies for consent.

1.43 If the landlord and the tenant do make an agreement on reasonableness, it will usually be valid. The landlord's duties under the LTA 1988 (to grant consent except when it can reasonably be withheld and subject only to reasonable conditions) will have effect subject to the agreement.

See 1.22 to 1.31 for when the LTA 1988 applies and details of the landlord's duties.

1.44 However, there are two important limitations on the agreement.

1.45 First, it must not provide for any circumstances or conditions to be framed by reference to any matter which will be determined by the landlord or by any other person, unless it also provides either:

- that the person's power to determine that matter must be exercised reasonably; or

- that the tenant has an unrestricted right to have any such determination reviewed by a person independent of both landlord and tenant whose identity is ascertainable by reference to the agreement, and the agreement provides for the determination made by any such independent person on the review to be conclusive as to the matter in question.

To the extent that the agreement falls foul of this, it will be of no effect.

Example

A 'new' tenancy provides that the landlord may withhold consent to an assignment if, in his opinion, the proposed assignee is not '*respectable and responsible*'. Because this leaves the matter in the hands of the landlord alone, the provision will be of no effect unless it goes on to require that:

- the landlord must act reasonably in assessing the character of the assignee; or
- the landlord's decision may be referred to a third party for review (for example, by adopting a procedure similar to the normal process for settling rent review disputes).

1.46 Secondly, if an agreement on reasonableness has the effect of excluding, modifying or frustrating the operation of the LT(C)A 1995, it will be void.

Example

A 'new' tenancy provides that the landlord may impose a condition on assignment that the tenant must enter into a guarantee of the tenant covenants for the full duration of the tenancy. The 1995 Act does not permit such a guarantee, and so this agreement on reasonableness will be void.

Legal advice should be taken if there is a question about this.

The difference between pre-conditions and agreements on reasonableness

1.47 Prior to the introduction of agreements on reasonableness, pre-conditions were the only valid way to avoid the general rule (that attempts to set out when consent may be withheld are invalid). It was therefore very important to understand whether any given clause would be interpreted as such an attempt, or as a pre-condition. This remains the case in situations where the second exception (assignments of non-residential new tenancies) does not apply.

1.48 Now, however, where agreements on reasonableness are allowed, it is less important to analyse the lease in that way. It is therefore less likely that the court will interpret a clause as imposing pre-conditions in those cases.[34]

1.49 However, the distinction can still sometimes be relevant. As explained at 1.22 to 1.31, in most cases where there is a fully qualified covenant against assigning, the LTA 1988 imposes an express duty on the landlord to consider applications within a reasonable time and to grant consent, unless it can be withheld reasonably. If the landlord fails to do so, he can be liable to pay damages to the tenant and it may also affect the landlord's ability to stop the assignment. It may therefore be relevant to decide whether a clause sets out a pre-condition or not if:

- a landlord delays in giving consent; or
- the landlord withholds consent for a different, unjustified, reason.

Example

A 'new' business tenancy contains a covenant 'not to assign except that subject to compliance with the following provisions of this clause the tenant may assign if it has obtained the landlord's written consent'. The following provisions state that consent will not be unreasonably withheld if the assignee is financially sound.

The tenant applies for consent to assign to an individual trader with no assets, heavy borrowings and a poor trading record. The landlord refuses consent on the basis that the assignee is a redhead and he only wants blondes in the premises. The tenant wants to claim damages.

- If the clause is a pre-condition, the tenant is unlikely to succeed (assuming the landlord takes the point in the proceedings!). Because the assignee was not financially sound, the landlord was under no obligation to deal with the application and the LTA 1988 will not apply. (In some cases, the tenant may be able to say that it is unfair for the landlord to be able to take the point at a late stage,[35] but on the facts given in the example, that is unlikely)

- If the clause is an agreement on reasonableness, then the landlord was under a duty to grant consent unless he could reasonably have withheld it. He is limited to the reasons he put in his response. Had he given lack of financial soundness as another reason, he would have withheld consent reasonably; as he gave only an unreasonable ground, he will be liable to pay damages.

Cases where the law does not transform the covenant

1.50   The general rule (that an attempt to set out when consent can be withheld is invalid) only applies where the law would transform a qualified covenant into a fully qualified one. In some cases, the lease expressly says that a covenant is fully qualified, but if it did not say so then the law would not transform it.

1.51   The most common types of covenant which are not transformed are covenants against changing use. See the table at 1.19 for a summary of which covenants are transformed, and Appendix 2 for full details.

1.52   In these cases, there is complete freedom to specify circumstances in which consent can be withheld, or conditions that the landlord can impose.

Whose consent is needed?

1.53   In most cases where consent is needed, it is the landlord who must consent. However, this is not always the case; sometimes the tenant needs someone else's consent as well or instead (for example, the superior landlord or a management company).

1.54   To find out whose consent is needed:

- First, look at the covenant that restricts what the tenant wants to do – this will often say whose consent is needed.

  If the covenant says that the consent of the 'Landlord' is needed, this will almost always mean the immediate landlord for the time being, i.e. the person the tenant pays the rent to. Usually a lease will expressly say that 'Landlord' means the named person who originally granted the lease *and his successors in title*, but even if these words are not included, the result will usually be the same. Similarly, if the covenant says that the consent of a superior landlord is needed, this will mean the person who now holds the freehold or other interest which the superior landlord held at the date of the lease.

- Secondly, look at the rest of the lease. Looking at the relevant covenant alone can be incomplete or misleading. Other parts of the lease, especially an interpretation section or a definitions section, will often add to or change the position.

Example

A sublease contains a covenant *'not to use the premises other than as a Chinese takeaway without the prior written consent of the landlord'*. The definitions section states that *'references to consent of the landlord shall be read as including the consent of the superior landlord'*. If the subtenant wants to sell Indian takeaway food instead, he will therefore need the consent of the superior landlord as well as that of the landlord.

- Thirdly, check any superior leases. A superior lease might prevent the subtenant's landlord from lawfully giving consent without his own landlord's consent. This could be done in a number of different ways, including:
  - an express covenant by the intermediate tenant not to consent to any application by the subtenant without the superior landlord's consent;
  - a covenant by the intermediate tenant that something *'shall not be done without landlord's consent'*, which will be done if the subtenant's application is granted;
  - a covenant by the intermediate tenant *'not to permit or suffer* [something] *to be done without landlord's consent'*, which will be done if the subtenant's application is granted; and
  - a covenant by the intermediate tenant that *'he, his successors in title and those deriving title from*

him' will not do something without his landlord's consent, which will be done if the subtenant's application is granted.

This list is not exhaustive and, if in doubt, legal advice should be obtained.

If the subtenant does not make sure that the superior landlord's consent is obtained, the superior landlord may be able to end the head-lease prematurely by forfeiting it. If that happens, then assuming the head-lease was granted before the subtenancy, the subtenancy ends automatically with the head-lease. The subtenant may or may not be able to get relief from forfeiture.

See 10.34 for more information about forfeiture by a superior landlord, and 10.30 to 10.32 on relief from forfeiture.

- Fourthly, consider if there are any other consents which may be needed. For example, is planning permission or listed building consent needed? If the lease is mortgaged, is the lender's consent needed? These considerations are outside the scope of this book.

Which covenants affect the tenant's plans?

Covenants in the tenant's lease

Covenants between landlord and tenant

1.55 The most important category of covenants which need to be checked are those in the tenant's lease. These do not only affect the original tenant and the original landlord who signed the lease when it was first granted. In the vast majority of cases, anyone who is the current landlord under the lease will be able to enforce a restriction relating to alienation, alterations or use, and anyone who is the current tenant under the lease, or who stands in the tenant's shoes, will be bound by the restriction.

1.56 It may not always be obvious who counts as the tenant, particularly in alienation cases. Some examples of people bound by the lease covenants are as follows:

- if the tenant has died and the lease has passed automatically to his executor or administrator, the executor or administrator is bound (at least if he wishes to dispose of the lease to a third party[36] and probably even if he wishes to pass it to a beneficiary);

- if the tenant has been made bankrupt and the lease has passed automatically to his trustee-in-bankruptcy, the trustee-in-bankruptcy is bound;[37]

- if someone else (for example, the liquidator of a tenant company) is acting on behalf of the tenant, he is bound;[38]

- if an authority has acquired the lease by compulsory purchase, it is bound;[39]

- if someone has a mortgage or other charge over a lease (a 'chargee'), he will usually be bound by the covenants in the lease. So if the lease is to be sold to pay off the tenant's debt, the chargee will have to comply with the covenants.

1.57 If there is unusual wording (for example, if the lease says that only certain categories of people will be bound and the current tenant is not within those categories) or if the lease was granted before 1926, specialist legal advice should be taken.

Covenants with other people

1.58 Sometimes the tenant will give covenants in the lease which can be enforced by other people. For example, a subtenant might covenant in the sublease that he will not assign his sublease or create a sub-underlease without the consent of both the landlord and the superior landlord. If the subtenant assigns or sublets without the superior landlord's consent, the superior landlord will be able to use some remedies, as described at 10.7, 10.14 to 10.15 and 10.33 to 10.34, even though the superior landlord was not named as a party to the sublease.[40] See also 1.53 to 1.54.

Covenants in licences or other deeds

1.59 If the tenant assigns or sublets, the landlord frequently insists on the assignee or the subtenant executing a formal deed in which they covenant directly with the landlord. These covenants usually repeat the covenants in the lease, but they may be more extensive. There may also be licences to make alterations or change use, or deeds of variation

1.60 Covenants in these ancillary documents will bind the person who signed the document. Sometimes they will be personal to that person, so if the lease is assigned on, the new tenant will not be bound by them. But sometimes, particularly if they are stated to vary the lease or be incorporated into the lease, future tenants will be bound. If in doubt, legal advice should be taken.

Other covenants

1.61 A tenant may be affected by other covenants which are not set out in his lease. The most common examples are as follows:

- Frequently a subtenant will covenant to comply with covenants in a superior lease. If he does so, he will be bound by these, and so will need to check the superior lease as well as his own lease before taking any action.

- Sometimes a subtenant will be bound by covenants in a superior lease even if he has not expressly promised to comply with them. The most common type of covenant of this kind is a restriction on use. This type of covenant affects anyone who is in occupation of the premises, whatever their status, if his right to be in the property can be traced back to the person who gave the covenant or even if he is a squatter.[41] The subtenant will usually be bound by this type of covenant, whether or not he actually knows about it.[42] A full discussion is beyond the scope of this book.
  However, the subtenant will not usually be affected by the alienation covenants which bind his own landlord, the tenant.[43]

Example

A tenant has a lease which contains a covenant 'not to assign, sublet or part with possession of the whole or any part of the premises without landlord's consent, such consent not to be unreasonably withheld'. He sublets the whole with his landlord's consent. The sublease contains no restrictions on alienation.

The subtenant later decides to assign the sublease. It can do so without needing the consent of either the tenant or the landlord.

- Sometimes a tenant will have given covenants to someone else. For example, if the lease is mortgaged there may be covenants in the mortgage deed.
- Sometimes there will be government covenants, such as agreements under s. 106 of the *Town and Country Planning Act* 1990, which affect the property.
- Sometimes the landlord will be bound by covenants which were given for the benefit of nearby property. This might be the freeholder covenanting with the neighbouring freeholder. Or it might be the landlord covenanting with another tenant of his. If so, the tenant may be bound by them even if he has not expressly covenanted to comply with them. Again, the most common type of covenant which will bind the tenant is one restricting use, and again this can affect anyone in occupation of the premises, sometimes even if the tenant does not know about it.

Example

A landlord let a number of shops in a parade to different tenants. He covenanted with one tenant, who ran a restaurant, that no other shop in the parade would be used as a restaurant. He later let another shop to a second tenant. The second tenant was deemed to know about the landlord's covenant with the first tenant, and was prevented from using its shop as a restaurant.[44]

A more detailed discussion of the circumstances in which a tenant may be bound by covenants which affect his landlord or superior landlords is beyond the scope of this book.

Who is responsible for a breach of covenant?

1.62 It might be thought that only the person who actually breaches the covenant can be liable for it. But in certain circumstances, other people may be liable as well. For example, a previous tenant may be liable for breaches committed by the current tenant or a tenant may be liable for breaches committed by his subtenant. The position will usually depend on the wording of the covenant.

1.63 This means that a landlord who is himself a tenant needs to check his own lease before granting consent, to ensure that the tenant's action will not make him liable to the superior landlord.

1.64 Liability for breaches of covenant also depends on whether the tenancy is a 'new' tenancy or not for the purposes of the LT(C)A 1995. A full analysis of who may be liable for a breach of covenant is outside the scope of this book.

How to analyse covenants

1.65 A full discussion of covenants, even of covenants which often require landlord's consent, is beyond the scope of this book. In complex cases it is usually wise to seek legal advice.

1.66 However, in modern leases, it is usually clear what the covenants are, and the key to analysing them is to read them carefully and sensibly. The basic principles are set out here, and there is more detailed guidance at 2.9 to 2.35, 3.10 to 3.26 and 4.16 to 4.48, but these alone may not always give the right answer.

Common sense

1.67 In the past, interpretation could be very legalistic, but now all documents, including leases, are interpreted by the same 'common sense principles by which any serious utterance would be interpreted in ordinary life'.[45]

1.68 Those principles have been summarised as follows:[46]

- Interpretation is the ascertainment of the meaning which the document would convey to a reasonable person having all the background knowledge which would reasonably have been available to the parties in the situation in which they were at the time of the contract.

- Subject to the requirement that it should have been reasonably available to the parties and to the exception to be mentioned next, the background includes absolutely anything which would have affected the way in which the language of the document would have been understood by a reasonable man.

- For reasons of practical policy, the law excludes from the admissible background the previous negotiations of the parties and their declarations of subjective intent, even though this would usually affect the way a reasonable man would understand the document.

- The meaning which a document (or any other utterance) would convey to a reasonable man is not the same thing as the meaning of its words. The meaning of words is a matter of dictionaries and grammars; the meaning of the document is what the parties using those words against the relevant background would reasonably have been understood to mean. The background may not merely enable the reasonable man to choose between the possible meanings of words which are ambiguous but even (as occasionally happens in ordinary life) to conclude that the parties must, for whatever reason, have used the wrong words or syntax.[47]

- The 'rule' that words should be given their 'natural and ordinary meaning' reflects the common sense proposition that we do not easily accept that people have made linguistic mistakes, particularly in formal documents. On the other hand, if one would nevertheless conclude from the background that something must have gone wrong with the language, the law does not require judges to attribute to the parties an intention which they plainly could not have had. In the words of one judge:

> 'If detailed semantic and syntactical analysis of words in a commercial contract is going to lead to a conclusion that flouts business common-sense, it must be made to yield to business commonsense'.[48]

1.69 It is important to note that the meaning of a document is the meaning which it conveys to reasonable readers with the parties' background knowledge. If one or both parties meant to say something different, that is irrelevant. It is possible to rectify a document, so that it does say something different, but this is very difficult to achieve. In this situation, evidence of previous negotiations is relevant.

Looking for the right words: reading the whole of the lease

1.70   Analysing restrictions should involve looking at the whole of the lease, not just the particular covenant which affects what the tenant wants to do. In particular, the definitions and interpretation sections of the lease should be checked, as these may give particular meanings to words.

Examples

(1) A tenant has a lease in which he covenanted '*not to assign the premises without landlord's consent*'. He has granted a lawful sublease, and the subtenant also covenanted '*not to assign the premises without landlord's consent*'. The subtenant applies to the tenant for consent to assign the sublease. At first sight, the tenant will not be in breach of his own lease if he grants consent: he did not covenant that he would not give consent to an assignment of a sublease without his own landlord's consent, and he will not be assigning the premises himself.

However, an earlier clause of the lease states that '*a covenant not to do something includes a covenant not to permit or suffer that thing to be done*'. The result of this is that the covenant means '*not to assign the premises or permit or suffer the premises to be assigned without landlord's consent*'. If the tenant gives consent to the assignment of the sublease, he will be permitting the premises to be assigned. So unless he gets his own landlord's consent, he will be in breach of his lease.

(2) Another tenant has a long lease of a house and wants to create a roof garden. The lease contains a covenant '*not to make any alterations to the demised premises, whether structural or otherwise, without landlord's consent, such consent not to be unreasonably withheld*'. The tenant is confident that there is no good reason to withhold consent to his plans.

However, the definitions section of the lease defines '*demised premises*' as '*all that terraced house known as 23 High Street, Markettown up to and including the plaster on the underside of the ceilings of the second floor but excluding the roofspace, roof and airspace above the roof*'. This means that the area where the tenant

> wants to make alterations is not included in the lease, and he has no right to make alterations, whether consent could reasonably be withheld or not.

See 3.3 on alterations to the 'demised premises'.

Literalism

1.71 Caution should be exercised before seeking to apply the words too literally if the outcome will make a nonsense of the covenant. A judge gave a graphic example of literalism, taken from the Works of William Paley:[49]

> 'The tyrant Temures promised the garrison of Sebastia that no blood would be shed if they surrendered to him. They surrendered. He shed no blood. He buried them all alive. This is literalism. If possible it should be resisted in the interpretative process.'[50]

Checklist: Identifying the need for consent

Use this checklist together with the checklists in chapter 2, 3 or 4 to decide whether the tenant can carry out his plans without consent, or (if he needs consent) whose consent he needs.

- Have you got and read the lease? Are there any restrictions that may affect what the tenant wants to do?
- Have you got and read any ancillary documents, such as a superior lease or a previous licence? Is the tenant affected by restrictions in those documents? See 1.59 to 1.61.
- Does the restriction affect what the tenant wants to do? See 1.65 to 1.71, and then 2.9 to 2.35, 3.10 to 3.26 or 4.16 to 4.48 for more detail.
- Is the meaning of the restriction affected by anything else in the lease, particularly the interpretation and definitions section?
- Does the restriction prohibit the act completely, or can it be done with consent? See 1.6 to 1.15.
- Does the lease expressly say that consent is not to be unreasonably withheld? If not, is this implied anyway? See 1.18 to 1.21, and then 2.79 to 2.91, 3.35 to 3.51 or 4.50 to 4.64 for more detail.
- Whose consent is needed? See 1.53 to 1.54.

2

Alienation

This chapter deals with alienation, the umbrella term used to cover all methods of disposing of a lease. The two most important methods are assignment and subletting.

The chapter explains:

- the tenant's position if there are no restrictions;
- the meaning of common lease clauses;
- when restrictions will be implied;
- when the landlord is prevented from demanding a premium;
- when the landlord must act reasonably; and
- how to overcome restrictions.

General principles

The starting point: the tenant is free to dispose of the premises

2.1   The general rule is that a tenant is free to dispose of his premises in any way.[51] However, this general rule is almost always displaced.

Tenancy not capable of being disposed of

2.2   Sometimes the tenancy is incapable of being disposed of. If the tenant tries to dispose of the lease, his attempt will be ineffective.

2.3   Tenancies at will are effectively incapable of being assigned or sublet. As soon as the landlord finds out about the assignment or subletting, the tenancy will come to an end.[52]

> A tenancy at will is a tenancy that can be terminated by either party at any time. It is most commonly found where the parties are negotiating for a new lease while the tenant is already in occupation, or is holding over after the expiry of his previous lease.

2.4   In addition, the law makes some residential tenancies incapable of being disposed of. This is explained at 2.40 to 2.54.

Restrictions

2.5   More commonly, the tenancy is capable of being disposed of, but there is a restriction. Most restrictions relating to alienation are express (written down in the lease, usually as a tenant's covenant) but in limited circumstances they can be implied (deemed to be in the lease even if not expressly stated). Implied restrictions are explained at 2.36 to 2.78. In the absence of a restriction of either type that relates to the proposed disposition, the tenant can go ahead.

2.6   If the tenant breaches a restriction, the disposition will be effective, but the landlord will be entitled to use the remedies explained in chapter 10.

Express restrictions

2.7   Restrictions in the lease are usually found in the tenant's covenants. They can be absolute (meaning a complete prohibition) or qualified (meaning that the landlord's consent is required). Many qualified covenants are fully qualified (meaning that they provide that consent cannot be withheld unreasonably). Covenants can also be hybrid: if a condition is not satisfied, the covenant is of one type (for example, absolute); if it is satisfied, the covenant is of another type (for example, fully qualified).

> For more information on the different types of covenant, and the importance of distinguishing between them, see 1.6 to 1.17.

2.8 Two points to note are:

- Most qualified alienation covenants are automatically transformed into fully qualified covenants: see 2.84 to 2.87 below. As explained at 1.16 to 1.17, fully qualified covenants offer much more protection to the tenant than qualified covenants.

- Many fully qualified alienation covenants offer even more protection than those relating to use or alterations. This is because the LTA 1988 requires landlords to deal with applications for consent to assign, sublet, charge or part with possession promptly and reasonably. It also gives tenants a right to claim damages if the landlord fails to deal with the application properly.

See 1.27 to 1.31 for details of when the LTA 1988 applies, 6.20 to 6.29 for the landlord's duties and 10.67 to 10.79 for the tenant's right to claim damages.

Understanding the restriction

2.9 Express restrictions come in various forms. It is crucial to be able to understand precisely what they mean: the tenant can do anything that is not covered by a specific restriction, without needing consent (provided it is not prohibited in any other way). Each covenant should be examined separately to see if the proposed transaction is restricted.

2.10 Interpreting words in leases is a difficult task and it is often wise to seek advice. Previous court decisions can be used as a guide to interpretation, but even a slight change in the words used or in the context may change the interpretation.

2.11 However, some general guidance on how to interpret covenants is given at 1.65 to 1.71. In addition, there are some key terms that appear in large numbers of leases, the meaning of which is relatively certain. These are explained below.

The meaning of common lease clauses

'Assign'

2.12 An assignment is a legal transfer of the whole of the unexpired term of the lease from the current tenant to someone else. For assignments of part, see 2.16 to 2.19.

2.13 'Assignment' includes all transfers effected by the tenant for the time being, for example:

- a transfer of the tenancy by an individual tenant into the name of a company controlled by him;
- a transfer of the tenancy by a company into the name of a group company;
- a transfer of the tenancy into the name of the tenant and another person jointly;
- a transfer of the tenancy from the joint names of two joint tenants into the name of one of them alone, at least if the transaction only involves the tenants;[53]
- a transfer of the tenancy from an assignee back to the original tenant;[54] and
- an assent by a deceased tenant's personal representatives, which vests the tenancy in a third party.[55]

2.14 'Assignment' does not usually include involuntary transfers,[56] for example:

- a transfer that the tenant is required to make under a valid compulsory purchase notice;[57]
- a transfer effected by a court order, such as an order vesting a lease in new trustees where the tenant, the old trustee, had disappeared;[58]
- the vesting of a lease in personal representatives after the tenant's death[59] or, if the tenancy is a joint tenancy, the passing of it to the survivor(s) when one of the joint tenants dies; or
- an automatic assignment to a trustee in bankruptcy under s. 306 of the *Insolvency Act* 1986, even where the tenant has petitioned for his own bankruptcy.[60] However, leases often contain express wording allowing the landlord to forfeit if the tenant becomes insolvent.

However, the position may be different if there is express or unusual wording. For example, *'doing or putting away with the term'* has been said to include involuntary assignments.[61]

2.15   Other points to note are as follows:

- *Formalities:* Except in unusual circumstances,[62] a legal assignment can only be effected by deed.[63] This is the case for all types of tenancy, even types that can be created orally. If the tenant attempts to assign the lease orally, or even if he puts it in writing without satisfying the formal requirements for a deed, there will be no breach of a covenant against assignment. There may, however, be a breach of some other covenant, for example a covenant against parting with possession.

> For information on what constitutes a deed, see the Glossary in Appendix 1.

Further, if the lease is already registered, the assignment will only take effect when the transfer is registered.[64] This means that the execution of a transfer deed will not be a breach of a covenant not to assign, but there will be a breach as soon as the transfer is registered. There may, however, be a breach of some other covenant, for example against holding on trust.

If the lease is not registered, there will be a breach of a covenant not to assign as soon as the transfer deed is completed, even if the lease needs to be registered as a result of the transfer.

- *Agreements to assign:* An agreement to assign a lease will not be a breach of a covenant against assignment.[65]
- *Invalid assignments:* If the assignment is invalid for any reason, it will not be a breach of a covenant against assignment.
- *Types of transaction prohibited:* A covenant *'not to assign'* only prohibits an assignment. It does *not* prohibit:

- subletting,[66] although unusual wording may extend the meaning to cover subletting as well, for example if the covenant is not to assign the premises *'for the whole or any part of the term'*;[67]
- granting an occupational licence;[68]
- granting a mortgage or other charge;
- making a declaration of trust;[69] or
- parting with possession of the premises in any way except a legal assignment.[70]

Assignments: part or whole?

2.16 Although an assignment must be a transfer of the whole of the unexpired term of the lease, and cannot relate to a shorter period of time, it is possible to assign a lease only in relation to part of the premises demised.

2.17 This frequently creates more problems than creating a sublease of part, and specialist legal advice should always be taken before assigning part of the premises.

Example

A tenant has a single lease of two adjoining shops. It wishes to downsize and dispose of one of the two shops. There are no covenants which prohibit assigning or subletting part of the premises, no alterations are required, and the buyer's proposed use is permitted. The tenant can either assign the whole of its interest in one of the shops, or it can grant a sublease of one of the shops. If it grants a sublease, it will be easier to control the use, state of repair and so on of the other shop than if it assigns.

2.18 Usually a lease will deal separately with assigning part of the premises and assigning the whole, so the position will be clear.

2.19 If it does not, then the wording of the lease may not necessarily produce the same result as the wording of the leases in previous court decisions. But the following principles are likely to be applied in most cases:

- If the restriction does not expressly refer to assignments of part of the premises or assignments of the whole of the premises, it will only prohibit assignments of the whole of the premises.[71]

- If the restriction prohibits assignments of *'any part'* of the premises, both assignments of part only and assignments of the whole of the premises will be prohibited.[72]

- If the restriction prohibits assignments of *'part only'* of the premises, only assignments of part and not assignments of the whole of the premises will be prohibited.

- If the restriction prohibits assignments of *'part'* of the premises, the position is unclear. We consider that this will only prohibit assignments of part and not assignments of the whole of the premises.

'Sublet' or 'underlet'

2.20   A sublease is the creation of a new lease by the tenant for a term shorter than the tenant's own lease. If the tenant attempts to create a sublease that will last for the whole of the unexpired term of his own lease or longer, the transaction will take effect as an assignment.[73] This is so even if it only affects part of the premises.[74] So there will be no breach of a covenant against subletting, but there will be a breach of a covenant against assigning.

Example

A tenant plans to sublet part of its office premises. Its own lease expires on 28 September 2015, so:

- a sublease that expires on or after 28 September 2015 will in fact constitute an assignment of the part (not a sublease); but
- a sublease expiring on or before 27 September 2015 will achieve its aim.

2.21   A covenant against subletting does *not* normally prohibit the following:

- assignments,[75] although unusual wording may extend the meaning to cover assignments as well, for example if the covenant prohibits sublettings for *'all or part of'* the term;[76]

- granting an occupational licence, such as taking in lodgers;[77]

- granting a charge over the premises;[78] or
- parting with possession of the premises in any way except a sublease.[79]

Sublettings: part or whole?

2.22 Usually a lease will deal separately with subletting part of the premises and subletting the whole, so the position will be clear.

2.23 If it does not, then the wording of the lease may not necessarily produce the same result as the wording of the leases in previous court decisions. But the following principles are likely to be applied in most cases:

- If the covenant does not expressly refer to sublettings of part of the premises or sublettings of the whole of the premises:
  - it will prohibit a single subletting of the whole of the premises; and
  - it will prohibit multiple sublettings of parts of the premises which result in the whole of the premises being sublet;[80] but
  - it will not prohibit one or more sublettings of part of the premises, provided the tenant retains part of the premises in his own possession.[81]

  This is because a covenant against subletting 'the premises' is only broken if 'the premises' are sublet, not if something smaller than 'the premises' is sublet. This principle can apply even if the covenant does not mention 'the premises', but only says *'not to sublet'*[82] or *'no subletting allowed'*.[83]

- If the restriction prohibits subletting *'any part'* of the premises, both sublettings of part only and sublettings of the whole of the premises will be prohibited.[84]
- If the restriction prohibits subletting *'part only'* of the premises, sublettings of the whole of the premises by a single subtenancy will not be prohibited.
- If the restriction prohibits subletting *'part'* of the premises, the position is unclear. We consider that this will only prohibit sublettings of part and not sublettings of the whole of the premises by a single subtenancy.

Examples

(1) A lease of a house contains no covenants restricting alterations or use, but does contain a covenant '*not to sublet the premises*'. The tenant converts the house into two flats, each directly accessible from the street. He sublets one of the flats and lives in the other himself. He is not in breach of the covenant against subletting.

(2) Subsequently the tenant decides to sublet the second flat as well. This means that the whole of the house is now sublet. He is now in breach of the covenant.

(3) Another tenant of a house with an identical lease converts it into two flats, which are accessed from a shared hallway. He sublets both the flats, with rights of access over the hallway which he does not sublet. He is unlikely to be in breach of covenant.

'Part with possession'

> For further help on the meaning of possession, see 'Leases and licences' in Appendix 1.

2.24 Possession is a complicated legal concept.

2.25 Particular points to note are:

- Possession is not the same as occupation; a person can be in occupation of premises without being in 'possession' of them.
- A company can be in possession of premises, although officers or employees of the company are the people physically occupying it.
- Where two joint tenants are in possession, if one of them abandons the property but the other remains, this is not a parting with possession.[85]
- A useful guide to help identify who is in possession of premises is to identify who has overall control of them.

2.26 A covenant against parting with possession will prohibit an assignment and a subletting,[86] even if there is no express wording referring to assignments or sublettings. It will also prohibit other forms of parting with possession, such as allowing prospective purchasers into possession before completion.[87]

2.27 It will not prohibit a declaration of trust unless the tenant also parts with possession by other means,[88] or a mortgage or other charge unless the circumstances are unusual. Nor will it prohibit the grant of an occupational licence, or sharing occupation.[89]

2.28 A covenant against parting with possession *'of the premises'* will not prohibit a parting with possession of part only, unless that is also expressly prohibited.[90] We consider that the position will be the same as that for assignments and sublettings, as described at 2.16 to 2.19 and 2.22 to 2.23.

'Part with the premises'

2.29 A covenant *'not to part with possession of the premises'* is different from a covenant *'not to part with the premises'*. The latter prohibits assignments, but not sublettings.[91]

'Share possession'

2.30 A covenant against sharing possession prohibits the conversion of a tenancy that was granted to a single tenant into what, in practical terms, will amount to a joint tenancy.[92] As with a covenant against *parting* with possession, 'possession' has a strict legal meaning; it does not mean 'occupation'.[93]

'Not to permit others to occupy or to share occupation'

2.31 A covenant against permitting others to occupy or against sharing occupation goes further than a covenant against parting with possession. It will prohibit occupational licences, as well as situations where the other person has 'possession' of the premises in a strict legal sense.

2.32 If the tenant is a limited company, it can only occupy through its human officers, employees and so on, so it will not be a breach of a covenant against sharing occupation if those people do occupy.[94]

2.33 Further, a covenant against sharing occupation will not be broken if a third party participates in the tenant's business. However, the covenant is likely to prohibit the carrying on of a separate business from that carried on by the tenant. There is no precise test to establish which of these two categories a particular situation falls into, so there will often be difficult questions of fact. Much will depend on what use is permitted under the lease.

Example

A tenant operated a nightclub by offering the club to external promoters, who staged club nights and who took full responsibility for admission revenues, advertising and administration. The promoters were not provided with keys to the premises, although they, and the artistes, were allowed into the premises approximately one hour before an event. The tenant remained exclusively responsible for everything concerned with the property as opposed to the event. The tenant had not shared occupation with the promoters.[95]

'Respectable and responsible person'

2.34 Responsible means 'able to discharge all obligations in respect of rent and covenants under the lease'. Respectable refers to the way in which a person conducts his business and his behaviour in the whole of his external relations. A company can be a respectable and responsible person.[96]

2.35 See examples 4 and 5 in 1.15 for details of when 'respectable and responsible' is and is not relevant.

Implied restrictions

2.36 In the case of residential tenancies, the law imposes some limitations on alienation.

2.37 These limitations fall into two categories:

- Some restrict the tenant's *ability* to dispose of the premises. If the tenant attempts to do so, his attempt will be ineffective.

- Others restrict the tenant's *right* to dispose of the premises. If the tenant does so anyway, the disposal will be effective, but the landlord will be able to use the remedies explained in chapter 10.

2.38 It is critical to identify correctly, at the outset, the type of tenancy that exists. There are some general guidelines in Appendix 1, but not a comprehensive description. This will sometimes be a task for a specialist lawyer, because of the large and complex variety of alternatives. Subject to that important caveat, the relevant principles are set out below.

2.39 These implied restrictions only affect secure, assured and Rent Act tenancies. If the tenancy is not residential, or you are confident that none of these types of tenancy is in issue, this section can safely be ignored.

Restrictions on the tenant's ability to alienate

1. Secure tenancies: assignment

> A secure tenancy is a type of public sector residential tenancy. See Appendix 1 for a fuller explanation.

2.40 Most secure tenancies are incapable of being assigned.[97] However, if the tenancy is a fixed term tenancy granted before 5 November 1982, there are restrictions on the tenant's *right* to assign, but not on his *ability* to assign. These are explained at 2.62 to 2.67. If the tenancy is a periodic tenancy, the tenant will be unable to assign it, even if it was granted before 5 November 1982.

2.41 The other exceptions are as follows:

- If the landlord is a co-operative housing association[98] the tenancy will be capable of being assigned in the same way as any other tenancy.[99]

- A secure tenant can exchange his tenancy with another secure tenant or with an assured tenant whose landlord is either the Housing Corporation, a registered social landlord or a housing trust which is a charity.[100] Particular points to note are:
  - The exchanges can involve two or more tenants; there need not be a direct swap.
  - All the tenants must have the written consent of their landlord to make the assignment.

– There are special provisions dealing with the grounds on which the landlord can withhold consent, the conditions which it can impose, how quickly the landlord must respond, and what form the landlord's response must take. These are dealt with at 6.30, 7.185 to 7.186 and 10.57 to 10.58.

- Secure tenancies can be assigned in pursuance of an order made in family proceedings under certain statutory provisions. These are set out in s. 91(3)(b) of the HA 1985, which is reproduced in Appendix 3.

- Secure tenancies can be assigned to a person who would be qualified to succeed if the tenant died immediately before the assignment.[101]

2.42 Even if one of these exceptions applies, the tenant should still check the lease carefully for any other restrictions.

- If the tenancy is being exchanged, the landlord will have to consent to the exchange in any event, so no further problem will arise.

- If the tenant wishes to assign pursuant to a court order, legal advice should be taken on whether the court has jurisdiction to make the order, and whether the tenant can argue that contractual restrictions can be overridden.

- If the tenancy is being assigned to a person who would be qualified to succeed, the assignment will be effective even if it is in breach of a restriction in the lease, but the breach may allow the landlord to terminate the tenancy.[102]

2.43 If the tenancy is not secure because the 'tenant condition' (that the tenant is an individual and occupies the dwelling-house as his only or principal home) is not satisfied and for no other reason, the tenancy remains incapable of being assigned unless one of the exceptions applies.[103]

2.44 These provisions only affect assignments of the whole of the dwelling-house. For assignments of part, see 2.72 to 2.75.

2.45 Similar provisions apply to introductory tenancies and demoted tenancies,[104] two other types of public sector tenancy. The exceptions are more limited.

2. Rent Act statutory tenancies: assignment

> A Rent Act tenancy is a type of private sector residential tenancy usually granted before 15 January 1989. It is 'protected' until the contractual term ends and can then become 'statutory'. See Appendix 1 for a fuller explanation.

2.46 A statutory tenancy is not really a tenancy at all; it is merely 'a personal right of irremovability'. This means that a statutory tenant has nothing which he can assign.[105] If he does attempt to assign the whole of the premises, and gives up residence, this will bring the statutory tenancy to an end. The landlord will therefore be able to obtain possession against the tenant or the assignee as appropriate.

2.47 However, a statutory tenant can transfer his status to another person if the landlord agrees.[106] The following points should be noted:

- The transfer must be effected by a written agreement.

- The incoming tenant need not be a statutory tenant prior to the transfer.

- The outgoing statutory tenant, the incoming statutory tenant and the landlord must always be parties to the agreement. In addition, if the consent of any superior landlord would have been required to an assignment of the previous contractual tenancy, the superior landlord must also be a party. If the landlord or the superior landlord is not a party, the agreement will have no effect.

- The incoming statutory tenant will take over the outgoing statutory tenant's status, and the outgoing statutory tenant will leave with nothing.

- It is an offence for any person to require any pecuniary consideration for entering into the agreement, although money can be demanded for specified items such as apportioned outgoings.

3. Rent Act statutory tenancies: subletting

2.48 The same logic would suggest that a statutory tenant is also incapable of subletting because he has no tenancy out of which he can create a subtenancy, but this is not the case.

2.49 It is clear that a statutory tenant can sublet part of the dwelling-house without giving up residence. Such a subtenancy will be valid.[107]

2.50 A subtenancy of part can last as long as the statutory tenancy does. It can even last longer than the statutory tenancy. This is because if the subtenancy is lawful (i.e. there was no contractual restriction against subletting in the protected tenancy or any breach has been waived[108]), then when the statutory tenancy is determined, the subtenant will be deemed to become the tenant of the landlord on the same terms as if the statutory tenancy had continued.[109]

2.51 However, if there was a contractual restriction and the breach has not been waived, the grant of the subtenancy will be unlawful. This entitles the landlord to apply for an order for possession.[110] If the order is granted, the subtenancy will end with the statutory tenancy.

2.52 It is unclear whether a statutory tenant can validly sublet the whole of the dwelling-house.[111]

4. Protected shorthold tenancies: assignment

2.53 A protected shorthold tenancy is a special type of Rent Act tenancy for a fixed term of between one and five years.[112] They were difficult to create, and specialist legal advice should be sought if there is a question as to whether one exists.

2.54 A protected shorthold tenancy is incapable of being assigned.[113] The only exception is that it can be assigned pursuant to a court order made in family proceedings under certain statutory provisions. These provisions are the same as those set out in s. 91(3)(b) of the HA 1985, which is reproduced in Appendix 3.

Restrictions on the tenant's right to alienate

2.55 As explained above, restrictions which affect the *right* (as opposed to the *ability*) to dispose of the property do not prevent dispositions from taking effect. They simply render the disposition unlawful, allowing the landlord to use the remedies explained in chapter 10.

1. Statutory periodic assured tenancies: assignment, subletting and parting with possession

> An assured tenancy is a type of private sector residential tenancy granted after 15 January 1989. See Appendix 1 for a fuller explanation.

2.56 If the tenancy is a statutory periodic tenancy (i.e. a periodic tenancy which arises at the end of a contractual fixed term), the law implies a qualified covenant against:

- assigning the tenancy (in whole or in part); or
- subletting or parting with possession of the whole or any part of the dwelling-house

(i.e. the tenant may only do so with landlord's consent).[114]

2.57 The implied term replaces any express term which was contained in the earlier contractual tenancy.[115]

2.58 It is important to note that there is no proviso that consent may not be withheld unreasonably, and the law does not imply one.[116]

2.59 The same term is implied into a periodic assured tenancy which arises on the termination of a long lease under Schedule 10 to the *Local Government and Housing Act* 1989. This Act gives security of tenure to certain long leaseholders.

2. Fixed term assured tenancies: assignment, subletting and parting with possession

2.60 If the tenancy is for a fixed term, the only restrictions on the tenant's right to assign are the restrictions set out in the lease. When the fixed term ends and is replaced by a statutory periodic tenancy, the qualified covenant described at 2.56 will be implied in place of the express term.

3. Contractual periodic assured tenancies: assignment, subletting and parting with possession

2.61 The qualified covenant described at 2.56 will be implied into a contractual periodic assured tenancy in certain circumstances:

- If the tenancy contains a provision relating to assignment, subletting or parting with possession, whether the provision prohibits it or permits it, and whether the provision is absolute or conditional, the terms of the tenancy will be the only terms that apply. The same will apply even if the provisions are not contained in the tenancy agreement itself.

- If the tenancy is a contractual periodic tenancy where a premium is required to be paid on the grant or renewal of the tenancy, then again the terms of the tenancy will be the only terms that apply. 'Premium' includes:
  – any fine or other like sum;
  – any other pecuniary consideration in addition to rent; and
  – any sum paid by way of deposit, other than one which does not exceed one-sixth of the annual rent payable under the tenancy immediately after the grant or renewal in question.

- Any other contractual periodic tenancy will have the qualified covenant described at 2.56 implied into it.[117]

4. Secure tenancies: assignment

> A secure tenancy is a type of public sector residential tenancy. See Appendix 1 for a fuller explanation.

2.62 Fixed term secure tenancies granted before 5 November 1982 are capable of being assigned, and the only restrictions on the tenant's right to assign are the restrictions set out in the lease.

2.63 However, if the tenant does assign his tenancy, the tenancy will no longer be secure.[118] This means that the statutory restrictions on the landlord recovering possession at the end of the term will be lifted, and the tenant loses the other benefits of having a secure tenancy.

2.64 The tenancy ceases to be secure whether the assignment is permitted by the terms of the tenancy or is in breach of it. Neither the tenant nor the assignee will be able to regain secure status under the tenancy. This means that the tenancy will not be continued when the contractual term comes to an end.

If the assignment is in breach of the terms of the tenancy, the landlord will usually be able to terminate the tenancy even before the end of the fixed term. The landlord can only do so if he has an express right to end the tenancy early if the tenant is in breach of covenant. The tenant can seek relief from forfeiture. This is explained at 10.16 to 10.32 and 10.39.

2.65 The same exceptions apply to this restriction on the tenant's *right* to assign as those which apply to the tenant's *ability* to assign, described at 2.41. If one of these exceptions applies, the tenant will be able to assign the tenancy without losing the secure status.

2.66 If the tenancy is not secure because the 'tenant condition' (that the tenant is an individual and occupies the dwelling-house as his only or principal home) is not satisfied and for no other reason, the position is the same as if it was secure, although the exceptions are more limited. As stated above, the tenancy cannot subsequently become secure.[119]

Example

A tenant of a council flat is not occupying it. This means that he is not a secure tenant. He assigns the tenancy to someone who will live there. The assignee can never become a secure tenant.

2.67 These provisions only affect assignments of the whole of the dwelling-house. For assignments of part, see 2.72 to 2.75.

5. Secure tenancies: subletting and parting with possession of the whole

2.68 The only restrictions on a secure tenant's right to sublet or part with possession of the whole of his dwelling-house are the restrictions set out in the lease.

2.69 However, as with assignments of fixed term secure tenancies granted before 5 November 1982, if the tenant does sublet or part with possession of the whole of the dwelling-house, the tenancy will cease to be secure and cannot subsequently become secure.[120] This rule applies whether the tenant sublets the whole of the dwelling-house in a single transaction, or whether he sublets first part of it and then the remainder. It also applies to all secure tenancies, whether fixed term or periodic, whenever they were granted.

2.70 We consider that as with assignments, the tenancy will cease to be secure whether the subletting or parting with possession is permitted by the terms of the tenancy or is in breach of them. This means that a fixed term tenancy will not be continued when the contractual term comes to an end, and a periodic tenancy can be terminated by the landlord at the end of the next period.

If the subletting or parting with possession is in breach of the terms of the tenancy, the landlord will usually be able to terminate the tenancy even before the end of the fixed term. The landlord can only do so if he has an express right to end the tenancy early if the tenant is in breach of covenant. The tenant can seek relief from forfeiture. This is explained at 10.16 to 10.32 and 10.39. If the tenancy comes to an end in one of these ways, this will automatically terminate the subtenancy as well.

2.71 If the tenancy is not secure because the 'tenant condition' (that the tenant is an individual and occupies the dwelling-house as his only or principal home) is not satisfied and for no other reason, the position is the same as if it was secure. As stated above, the tenancy cannot subsequently become secure.[121]

Example

A tenant of a council flat is not occupying it. This means that he is not a secure tenant. He sublets to someone who will live there. The subtenancy comes to an end and the tenant moves back into the dwelling-house. It is still not secure and can never become secure.

6. Secure tenancies: subletting and parting with possession of part

2.72　It is a term of every secure tenancy that the tenant will not, without the written consent of the landlord, sublet or part with possession of part of the dwelling-house.[122] Consent may not be unreasonably withheld.[123]

2.73　This provision only restricts the creation of a sublease and parting with possession. It will not apply to an occupational licence, and specific provision is made to allow secure tenants to take in lodgers: see 2.76. However, as an assignment counts as parting with possession (see 2.26), this provision will also restrict assignments of part of the dwelling-house.

> For further information on the difference between a lease and a licence, see Appendix 1.

2.74　There are special provisions relating to grounds on which the landlord can refuse consent, whether conditions can be imposed, and what form the landlord's response should take. These are explained at 6.31 and 7.187 to 7.188.

2.75　If the secure tenant does grant a sublease of part, the sublease will be valid, but the tenant will be in breach of the lease. This means that the landlord will be able to terminate the tenancy if it is reasonable to do so.[124] If the tenancy is terminated, the subtenancy will also be terminated automatically.

7. Secure tenancies: taking in lodgers

2.76　It is a term of every secure tenancy that the tenant may allow any persons to reside as lodgers in the dwelling-house.[125]

8. Rent Act tenancies: assignment and subletting of the whole

> A Rent Act tenancy is a type of private sector residential tenancy usually granted before 15 January 1989. It is 'protected' until the contractual term ends and then becomes 'statutory'. See Appendix 1 for a fuller explanation.

2.77 No additional restrictions as such are implied into a protected or statutory tenancy. So if the tenant assigns or sublets during the protected tenancy, the landlord will not be able to use any of the remedies described in chapter 10, unless there is an express term restricting assignment or subletting in the lease.

2.78 However, if the tenant assigns or sublets the whole of the premises, when the protected tenancy comes to an end and is replaced by a statutory tenancy, the landlord may be able to obtain a court order for possession. This is the case whether or not there is an express restriction in the lease. This forms an indirect restriction on assignment and subletting. For further details, see 10.43 to 10.46.

Additional tenant protection

2.79 In some circumstances the law gives the tenant additional protection. This comes in two forms:

- Sometimes the law prevents the landlord from demanding a premium.
- Sometimes the law imposes a requirement for the landlord to act reasonably.

The position depends on whether the wording of the restriction is absolute, qualified or fully qualified.

> For an explanation of absolute, qualified and fully qualified covenants, see 1.6 to 1.15. For more information on what constitutes a premium, see 8.19 to 8.22.

Absolute restrictions

Premiums

2.80 In cases where the disposal is absolutely prohibited, if the tenant asks for consent the landlord is entitled to ask for a premium as the price of his consent.[126]

Acting reasonably

2.81 Absolute covenants arc not made fully qualified. They always constitute a complete prohibition unless they can be overcome in one of the ways described at 1.9.

Qualified restrictions

Premiums

2.82 Where there is a qualified restriction against alienation, s. 144 of the LPA 1925 usually prohibits the landlord from demanding a premium.

The prohibition does not prevent the landlord (or RTM company – see 5.23 to 5.25) from requiring the payment of a reasonable sum in respect of any legal or other expense incurred in relation to the licence or consent. The types of costs the landlord can recover are dealt with in chapter 8.

2.83 The prohibition does not apply if the lease expressly excludes the section or specifically states that the landlord may demand a premium. Even then, legal advice should be taken before demanding a premium, particularly in the case of 'new' tenancies.

However, as explained in the next paragraph, almost all qualified restrictions are automatically transformed into fully qualified restrictions, and except in unusual circumstances, a demand for a premium will be unreasonable: see 2.88 to 2.90. This means that s. 144 will usually be relevant only to limited types of tenancy where the transformation does not apply, such as statutory periodic assured tenancies.

> See Appendix 2 for details of when s. 144 does not apply, and when qualified covenants are not made fully qualified.

Acting reasonably

2.84 If the lease contains a qualified covenant against alienation, s. 19 of the LTA 1927 usually transforms it so that consent cannot be withheld unreasonably.[127]

Example

A tenant has a lease which contains a covenant *'not to assign, underlet, or part with possession of the premises without the landlord's prior written consent'.*

This is a qualified covenant: the tenant may assign, underlet or part with possession, but only with the landlord's prior consent. However, the law makes it fully qualified: the landlord's consent is not to be withheld unreasonably.

2.85 It is important to note the following:

- The transformation only applies to four types of alienation: assigning, underletting, charging or parting with the possession. Covenants that relate to a different type of alienation (such as a declaration of trust or sharing occupation) are unaffected.

- The transformation does not apply to all qualified covenants. The most common exception is when a qualified covenant is implied into an assured tenancy: see 2.56 to 2.61.

See Appendix 2 for full details of when the transformation applies.

2.86 As explained at 1.33 to 1.35, where the transformation applies, it is usually impossible to 'contract out'. This means that if a lease states that the transformation does not apply, or tries to set out circumstances in which it will be reasonable to withhold consent, or conditions which can reasonably be imposed, these provisions of the lease will usually be invalid.

2.87 There are two important exceptions to this rule:

- The provisions might be interpreted as setting out conditions which must be satisfied before the tenant can make an application for consent: see 1.36 to 1.40.

- 'New' tenancies that are not residential tenancies may set out circumstances in which it will be reasonable to withhold consent, or conditions which can reasonably be imposed. This only applies in the case of assignments: see 1.41 to 1.49.

Most tenancies granted in or after 1996 are 'new' tenancies. See Appendix I for more details.

Fully qualified restrictions

Premiums

2.88   In cases where the landlord may not withhold consent to alienation unreasonably, it is usually impermissible for him to ask for a premium for giving consent. If he does, he will usually be regarded as having withheld consent unreasonably, so that the tenant can go ahead with the transaction.[128]

2.89   The landlord (or RTM company – see 5.23 to 5.25) is, however, entitled to ask for payment of 'a reasonable sum in respect of any legal or other expenses incurred in connection with such licence or consent'.[129] The types of costs the landlord can recover are dealt with in chapter 8.

2.90   In unusual cases (where there is a pre-condition or a valid agreement on reasonableness stating that the landlord may demand a premium) we consider that a premium may be demanded. However, legal advice should be taken first. Some commentators consider that such a provision would be void in the case of a 'new' tenancy, at least if it is an agreement on reasonableness and not a pre-condition.[130]

For information on pre-conditions and agreements on reasonableness, see 1.36 to 1.49.

Acting reasonably

2.91   In the case of fully qualified restrictions, the landlord is already required to act reasonably. However, if the modification described at 2.84 applies and is more favourable to the tenant than the wording of the lease, the tenant can rely on the modification.[131]

Overcoming restrictions

2.92 If there is a qualified covenant (i.e. the tenant can alienate with the landlord's consent) then the law may help tenants. The most important examples are:

- qualified covenants become fully qualified; and
- qualified covenants in long building leases have no effect.

Qualified covenants become fully qualified

2.93 If the lease contains a qualified covenant against alienation, the law gives a helping hand: see 2.84 to 2.87.

Qualified covenants in long building leases have no effect

2.94 If certain conditions are satisfied, the tenant can assign, sublet, charge or part with possession freely, without needing consent, even though the lease states that he does need consent.[132]

2.95 This rule:

- only applies where there is a qualified covenant (it does not modify absolute covenants); and
- does not apply to certain types of lease or covenant. See A2.6 to A2.10 in Appendix 2 for details of which types of lease and covenant are and are not affected.

2.96 The conditions are:

- The lease must be for more than 40 years.

  It is likely that this means that the lease must last for at least 40 years from the later of the date of grant and the date of commencement.[133] If so, a lease granted on 1 January 1980 for a term of 45 years from 1 January 1980 would qualify, but a lease granted on 1 January 1990 for a term of 45 years from 1 January 1980 would not qualify.

- The lease must have been made in consideration wholly or partially of the erection, or the substantial improvement, addition or alteration of buildings.

- The landlord must not be a Government department or local or public authority, or a statutory or public utility company.[134]

- The lease must not be a mining lease.[135]

- The transaction may be an assignment, a subletting, a charge or a parting with possession and it may be effected either by the tenant or any subtenant, whether immediate or not.

- If the transaction is an assignment (including parting with possession on assignment[136]), the lease must either be an 'old' tenancy or, if it is a 'new' tenancy, it must be a residential lease. A residential lease is a lease by which a building or part of a building is let wholly or mainly as a single private residence.[137]

> Most tenancies granted in or after 1996 are 'new' tenancies. Other tenancies are usually called 'old' tenancies. See Appendix 1 for more details.

- The transaction must be effected more than seven years before the end of the term.

- Notice in writing of the transaction must be given to the landlord (or RTM company – see 5.23 to 5.25) within six months after the transaction is effected.

2.97 If the conditions are satisfied, then the tenant, subtenant, sub-undertenant and so on can assign, sublet, charge or part with possession without needing consent.

2.98 However, this only affects the need for consent to assign, sublet, charge or part with possession. It does not affect any other covenants that the tenant is bound by which affect the proposed transaction. So, for example, a separate covenant to obtain a guarantor on any assignment is not affected.[138]

2.99 If the conditions are not satisfied, qualified covenants may still be made fully qualified, as explained at 2.84 to 2.87: although the tenant will need consent, consent may not be withheld unreasonably.

Example

In January 1950, a tenant took a 65-year lease, covenanting to build a hotel on the property in consideration of the grant of the lease. The lease contained a covenant not to assign, sublet or part with possession of the premises without the consent of the landlord. It also contained a covenant not to sublet the premises except on terms that the subtenant covenanted not to assign, sublet or part with possession of the premises without the consent of the superior landlord.

When the hotel was built, the tenant granted a sublease to its subsidiary, which operated the hotel. Because the head-lease was a long building lease, the tenant did not need landlord's consent to sublet. It did, however, need to include a covenant in the sublease not to assign, sublet or part with possession of the premises without the consent of the superior landlord, and it did need to give the landlord written notification within six months of the subletting.

In 2006 the tenant and subtenant decided to sell all their hotel interests. They quickly found a buyer for the sublease. Because the head-lease was a long building lease, the subsidiary could assign the sublease without superior landlord's consent, but again it had to give written notification to the superior landlord.

A buyer for the head-lease was only found in late 2008. By this time the head-lease would expire in less than seven years. This meant that the modification for long building leases did not apply, so the tenant did need landlord's consent to assign the head-lease. However, the restriction on assignments was made fully qualified, so the landlord could not withhold consent unreasonably.

Implied modification of a covenant

2.100 In exceptional circumstances, the court may imply a term that makes it easier for the tenant to assign. These cases are likely to be very rare, and legal advice should be taken.

Example

A landlord was entitled to require guarantees from two or more directors of any assignee that was a private company, such directors being of suitable standing as approved by the landlord. There was no express proviso that consent would not be withheld unreasonably, nor did the law imply one. While the judge did not go so far as to imply that type of proviso, he did imply a term that the landlord had to act reasonably in exercising his right to require guarantees.[139]

Checklist: Alienation

Use this checklist to decide whether an intended transaction is restricted by the lease, and if so how the restriction operates and whether it can be avoided.

- Is there an express restriction in the lease? If so, does it cover the intended transaction? See 2.9 to 2.35.
- Does the restriction prohibit the transaction completely, or can it be done with landlord's consent? See 1.6 to 1.15.
- Does the lease expressly say that consent is not to be unreasonably withheld? If not, is this implied anyway? See 2.84 to 2.87.
- If there is no express restriction, is a restriction implied? If so, does it cover the intended transaction? See 2.36 to 2.78.
- If consent cannot be unreasonably withheld, tenants should consider how to make a proper application for consent (see chapter 5), and landlords should consider how to respond (see chapters 6 and 7).
- If the restriction prohibits the transaction completely, or if consent can be withheld, can the restriction be avoided? See 2.92 to 2.100.

3

Alterations

This chapter explains:

- the tenant's position if there are no restrictions;
- the meaning of common lease clauses;
- when restrictions will be implied;
- when the landlord is prevented from demanding a premium;
- when the landlord must act reasonably; and
- how to overcome restrictions.

General principles

The starting point: the tenant is free to alter the demised premises

3.1　A tenant is free to alter leasehold premises in any way, unless there is a restriction. Restrictions can be either express (written down in the lease, usually as a tenant's covenant) or implied (deemed to be in the lease even if not expressly stated: see 3.27 to 3.34). In the absence of a restriction of either type that relates to the tenant's intended works, the alteration can go ahead.

3.2　If the tenant breaches a restriction, the landlord will be entitled to use the remedies explained in chapter 10.

The demised premises

3.3　A tenant is only entitled to alter the *demised* premises. A surveyor should therefore begin by checking whether a tenant's proposed works will fall outside the demise, even partially. This means carefully checking the lease plan and the description of the demised premises to see exactly what is leased to the tenant. For example, is the whole of a wall included in the demise, or only half its width, or only the internal finish? Works or alterations going beyond the demised premises are prohibited,[110] and are a trespass against the other property (unless the tenant has separate rights over it).

Access to other land

3.4 Even if the alterations are entirely within the demised premises, a tenant still cannot access or use other property while doing the works without permission from the owner. This means that if the tenant needs to walk onto, or if the alterations will be attached to, other property, the works might be impossible to carry out regardless of the terms of the lease.

3.5 Surveyors may therefore need to check whether the tenant has rights over any other property. Check the tenant's deeds for any rights of access previously granted by neighbours, and look at the lease. Leases often grant rights over the landlord's nearby property. For example, a tenant of shop premises who wants to install an air conditioning unit may need access over a rear yard. If the yard is owned by the landlord, the lease may grant rights of way across it.

3.6 In some very limited cases the law may help the tenant to get access (see 3.100 to 3.103).

General restrictions

3.7 The tenant's right to make alterations may also be restricted by the general law, most notably by planning legislation.

Express restrictions

3.8 Restrictions in the lease are usually found in the tenant's covenants. They can be absolute (meaning a complete prohibition) or qualified (meaning the landlord's consent is required). Many qualified covenants are fully qualified (meaning that they provide that consent cannot be withheld unreasonably). Covenants can also be hybrid: if a condition is not satisfied, the covenant is of one type (for example, absolute); if it is satisfied, the covenant is of another type (for example, fully qualified).

For more information on the different types of covenant, and the importance of distinguishing between them, see 1.6 to 1.17.

3.9 Most qualified alterations covenants are automatically transformed into fully qualified covenants: see 3.40 to 3.44. As explained at 1.16 to 1.17, fully qualified covenants offer much more protection to the tenant than qualified covenants.

Understanding the restriction

3.10 Express restrictions come in various forms. It is crucial to be able to understand precisely what they mean, because any type of alteration not covered by a specific restriction may be carried out without consent (provided it is not restricted in any other way). Each covenant should be examined separately to see if the proposed works are restricted.

3.11 If the tenant plans to carry out various alterations together, each individual item in the proposed works must be looked at separately, to see whether it will breach a restriction. If some alterations are restricted and some are not, the tenant is free to split the tasks and begin the unrestricted alterations immediately, if he wishes.

3.12 The purpose for which the lease was granted is often relevant to interpreting the restriction. The courts are likely to construe an alterations clause to allow the tenant to do 'those acts which are convenient and usual for a tradesman to do in the ordinary conduct of his business'.[141] For example, the clock referred to in 3.15 below was permitted partly because it was incidental to the tenant's business.

3.13 Interpreting words in leases is a difficult task and it is often wise to seek advice. Previous court decisions can be used as a guide to interpretation, but only where the words of the lease and all the circumstances are the same.

3.14 However, some general guidance on how to interpret covenants is given at 1.65 to 1.71. In addition, there are some key terms that appear in large numbers of leases, the meaning of which is relatively certain. These are explained below.

The meaning of common lease clauses

'Alterations'

3.15 A covenant simply not to make alterations to the demised premises is quite limited in extent. In this context 'alteration' is likely to mean only an alteration that affects the form or structure of a building.

> *Example*
>
> Attaching a large clock to the outside of a watchmaker's premises using six-inch iron bolts was not an 'alteration' since the holes in the wall were held not to be a change in the form or structure of the building.[142]

'Elevation or architectural decoration'

3.16 Similarly, a restriction on alterations of this type has been held to apply only to the fabric of the building.

> *Example*
>
> A landlord complained about a temporary illuminated framework erected by the tenant to advertise whisky, ginger ale and Bovril. The court decided the restriction did not prohibit an alteration in appearance caused by 'temporary advertisements and frameworks which can be removed at any time, leaving the structure the same as before'.[143]

'External appearance'

3.17 A covenant not to alter the external appearance of a building is more wide-ranging.

Examples

(1) Letting a wall as a bill-posting station was an alteration of the external appearance.[144]

(2) Replacing air conditioning plant amongst a 'hideous jumble of backs of buildings' was not.[145]

'Cutting or maiming principal walls or timbers'

3.18 This is a common form of restriction in older leases. Despite the language of the clause, there is no need for any destructive element.

Examples

(1) This clause has been held to restrict the installation of an advertising station on the façade of a building, because it would cut into the wall.[146]

(2) Making notches in joists in order to install air conditioning was also a breach.[147]

'Structural alterations'

3.19 'Structural' alterations are alterations that involve the fabric of the building.[148] It is usually necessary to take expert advice about whether works will affect 'structural' parts.

Examples

(1) Installation of a modern central heating system has been held to be a structural alteration.[149]

(2) Windows have been held not to be part of the structure, so replacing steel window frames with aluminium was not a structural alteration.[150]

However, in other circumstances, windows might be part of the structure and central heating might be non-structural.

'Internal demountable partitioning'

3.20 It is common for office leases to permit tenants to install this type of partitioning. This allows the tenant to divide up open plan space and avoids burdening the landlord with close management of minor alterations.

Impact of other covenants

3.21 Other covenants can sometimes form an indirect restriction on alterations.

Compliance with statutes

3.22 Most well-drafted leases require tenants to comply with any statutes or regulations affecting the use of the premises. This will sometimes impose an indirect restriction on alterations that would otherwise be permitted (for example, an alteration that would put the property in breach of fire regulations).

Not to apply for planning permission without consent

3.23 Many leases restrict the tenant's right to apply for planning permission for an alteration without the landlord's consent. This presents another potential restriction for the tenant if the works would require planning approval. Although the lease may not restrict the works as such, if the tenant cannot do them without planning consent, he must get the landlord's approval before applying.

3.24 The landlord need not act reasonably, unless there is express wording in the lease requiring him to do so.[151] He may also demand a premium in return for his consent. Such a clause may therefore enable a landlord to prevent works unreasonably, even if his consent to the works themselves may not be unreasonably withheld.

The repairing covenant

3.25 It is possible for alterations to be prohibited by a standard repairing covenant. An obligation to repair and maintain the premises includes an obligation not to destroy them. Therefore, even in the absence of a restrictive alterations clause, a tenant might be prohibited from doing certain works, especially of a destructive nature.

3.26 However, much will depend on the purpose for which the lease was granted. If it appears that alterations were intended to be permitted, the repairing covenant will not override that.

Examples

(1) Taking down a brick wall breached the covenant to repair and maintain it.[152]

(2) In a case of a lease for 1,000 years with no express restriction on alterations, it was held that demolishing and redeveloping was not prohibited.[153] The court felt that, in a lease of that length, alterations must have been intended.

Implied restrictions

3.27 Even where a lease says nothing that might restrict alterations, it is not safe to assume that the tenant is entirely free to alter the premises. The law imposes some restrictions in any case. The most important are:

- the law of 'waste';
- restrictions affecting public sector residential leases; and
- restrictions affecting some Rent Act tenancies.

Waste

3.28 Waste is a complicated legal concept, but, in simple terms, it can make it unlawful for a tenant to change the use or character of the property.

Examples

(1) A tenant of a shop and house pulled down a partition and changed the premises to a shop only.[154]

(2) A tenant failed to repair damage caused when removing fixtures and fittings at the end of the term.[155]

In both of these cases, the tenant's actions amounted to waste.

3.29 However, provided that the change of use of the premises is permitted by the lease, the concept of waste is unlikely to apply.

Example

When a tenant converted an old chapel into a cinema with various alterations to the premises, this was not waste because using the premises as a cinema was allowed.[156]

3.30 Assessing whether waste applies is likely to require specialist legal advice. If it does, the landlord may be entitled to stop the alterations.

Secure tenancies

A secure tenancy is a type of residential public sector tenancy. For more details, see Appendix 1.

3.31 Secure tenancies (other than those where the landlord is a co-operative housing association[157]) are deemed to contain an implied term that alterations or additions cannot be carried out without the landlord's consent.[158] The restriction includes:

- additions and alterations to the landlord's fixtures and fittings;
- additions and alterations connected with the provision of services to the premises;
- the erection of a wireless or television aerial; and
- external decoration.

The landlord's consent is not to be withheld unreasonably.[159] The landlord's duties and factors relevant to reasonableness are explained at 6.42 to 6.43 and 7.189 to 7.190.

3.32 The tenant may be entitled to compensation,[160] and there are limits on the landlord's ability to increase the rent after an improvement.[161]

Rent Act tenancies

A Rent Act tenancy is a type of private sector residential tenancy usually granted before 15 January 1989. See Appendix 1 for more details.

3.33 Protected and statutory tenancies under the RA 1977 have an implied term similar to secure tenancies, described in the previous section. The term is not implied if the tenant has been given notice:

- that the tenancy was to be a protected shorthold tenancy; or
- seeking possession on certain grounds (unless the tenant proves that when the landlord gave the notice it was unreasonable for the landlord to expect to recover possession on the grounds stated in the notice).[162]

3.34 Specialist legal advice may be required to confirm whether a protected or statutory tenancy exists, and if so whether the circumstances are such that the restriction will be implied.

Additional tenant protection

3.35 In some circumstances the law gives the tenant additional protection. This comes in two forms:

- Sometimes the law prevents the landlord from demanding a premium.
- Sometimes the law imposes a requirement for the landlord to act reasonably.

The position depends on whether the wording of the restriction is absolute, qualified or fully qualified.

For an explanation of absolute, qualified and fully qualified covenants, see 1.6 to 1.15. For more information on what constitutes a premium, see 8.19 to 8.22.

Absolute restrictions

3.36 A business tenant may be able to overcome absolute restrictions: see 3.76 to 3.93. If he cannot, the following principles apply.

Premiums

3.37 In cases where the alterations are absolutely prohibited, if the tenant asks for consent the landlord is entitled to ask for a premium as the price of his consent.[163]

Acting reasonably

3.38 Absolute covenants are not made fully qualified. They always constitute a complete prohibition unless they can be overcome in one of the ways described at 1.9.

Qualified restrictions

Premiums

3.39 The position is the same for qualified restrictions as for absolute ones. Hence, the landlord may ask for a premium for giving consent. However, there will be very few such cases because, as explained in the next paragraph, the law transforms most qualified alterations covenants into fully qualified covenants.

Acting reasonably

3.40 Section 19 of the LTA 1927 usually transforms a qualified covenant so that, if the tenant wishes to carry out *improvements* to the demised premises, consent cannot be withheld unreasonably.[164]

> See Appendix 2 for full details of which covenants are not automatically transformed.

3.41 Where the transformation applies, it is impossible to 'contract out'. Therefore an attempt in the lease to avoid the transformation (whether deliberate or not) will usually have no effect. See 1.33 to 1.35 for further details. However, there is an important exception: the provisions might be interpreted as setting out conditions which must be satisfied before the tenant can make an application for consent: see 1.36 to 1.40.

3.42 It does not matter whether or not the covenant refers to *'improvements'* expressly (in practice most leases do not), provided it has the effect of restricting improvements. Therefore, a normal covenant against carrying out any alterations without the landlord's consent would have the effect that consent could not be withheld unreasonably in the case of improvements.

3.43 The law does not expressly define the word improvements. However, the meaning is judged from the tenant's point of view. If the works will improve the tenant's use of the premises, or make them 'more convenient and comfortable to him',[165] that will suffice, even if the landlord might not regard the works as improvements. Since this is, of course, the motivation behind most alterations carried out by tenants, the transformation applies in a wide variety of circumstances.

Examples

(1) A tenant had a lease of a residential flat which formed the top half of a house. The tenant wanted to convert the roof space above the flat (which was included in the demise) into a living area. This involved the installation of dormer windows which would change the appearance of the roof. The covenant in the lease was:

> 'Not at any time during the said term without the licence in writing of the Lessor first obtained to ... make any alteration in the plan or elevation of the demised premises or in any of the party walls or the principal or load-bearing walls or timbers thereof'.

The court held that the works were improvements and that the transformation applied. Therefore, the landlord's consent could not be withheld unreasonably. The works were allowed to proceed.[166]

(2) A tenant of shop premises wanted to create a larger unit spanning the demised property and other land to the rear (which was leased from another landlord). The rear wall of the premises would be pulled down, and the main staircase and staff accommodation removed. Even despite the obviously negative impact of the works on the landlord's interests, the court held that they were 'improvements' because they helped the tenant get the most out of the premises.[167]

(3) The courts have also decided that:

- making an opening between the demised premises and other premises;[168] and
- moving a staircase[169]

qualified as improvements.

3.44 A tenant cannot install equipment that trespasses on any land or property retained by the landlord. So, for example, works that involved attaching a flue for a hot water system to a balcony retained by the landlord could not be 'improvements'.[170] An improvement may, however, involve work to the demised premises together with other land owned by the tenant.[171]

Fully qualified restrictions

Premiums

3.45 In cases where the landlord may not withhold consent to alterations unreasonably, he usually cannot ask for a premium for giving consent. If he does, he will usually be regarded as having withheld consent unreasonably, so that the tenant can go ahead with the works.[172]

3.46 However, at least in the case of improvements, the landlord may be entitled to compensation.[173] If the works would reduce the value of the premises, or any of the landlord's neighbouring premises, the landlord is entitled to payment of reasonable compensation as a condition of giving consent. If there is an RTM company (see 5.23 to 5.25), the RTM company has no right to compensation.[174]

3.47 The way in which the landlord expresses his request for compensation is important:

- If the landlord said, 'I give my consent provided you pay me £x (an unreasonable amount) and I will not take a penny less', consent would probably be found to have been withheld unreasonably and the tenant will be free to go ahead.
- If the landlord said, 'I give my consent provided you pay a reasonable sum, and I am advised and consider that £x is such a sum', he would probably be found to have acted reasonably.[175]

It is sensible for the landlord to put forward an amount that he considers reasonable, but state that he will accept such other sum as may be found to be reasonable by the court.

3.48 The courts have held that if loss of value is the landlord's only objection to the works, consent cannot be refused.[176] Instead, the landlord must consent subject to receiving appropriate compensation.

> See chapter 7 for more information on grounds of objection and conditions that the landlord may impose.

3.49 There is express permission to ask for payment of 'any legal or other expenses properly incurred in connection with such licence or consent'.[177] The types of costs which can be recovered are dealt with in chapter 8.

3.50 In unusual cases (where there is a pre-condition stating that a premium may be demanded or if the alteration is not an improvement), the landlord may be able to demand a premium. Legal advice should be obtained first.

> See 1.36 to 1.40 for more information about pre-conditions, and 1.50 to 1.52 on the position where the alteration is not an improvement.

Acting reasonably

3.51 In the case of fully qualified restrictions, the landlord is already required to act reasonably. However, if the modification described in 3.40 applies and is more favourable to the tenant than the wording of the lease, the tenant can rely on the modification.[178]

Overcoming restrictions

3.52 In certain circumstances the law helps tenants to overcome restrictions. This often applies even where the restriction is absolute (i.e. a complete prohibition). This can, therefore, be very valuable for a tenant.

3.53 The most important ways of overcoming restrictions relate to:

- compliance with disability laws;
- electronic communications equipment;
- improvements to properties let on business tenancies; and
- converting dwellings into flats.

These, and various others, are explained below. The key points of each are summarised in a table at the end of this section.

3.54 The ways in which the law helps fall into two categories:

- automatic changes to restrictions (where the law gives the lease a different meaning); and
- discretionary changes to the lease (where the tenant must apply to court, which *may* change the restriction).

Automatic change 1. Improvements: qualified covenants become fully qualified

3.55 If the lease contains a qualified covenant against alterations, the law gives a helping hand: see 3.40 to 3.45.

Automatic change 2. Compliance with disability laws

3.56 The law makes it easier for tenants to carry out works in order to comply with disability laws. These laws fall into two categories:

- the duty on employers to make reasonable adjustments to premises;[179] and
- the duty on service providers not to discriminate against disabled customers.[180]

3.57 Any restriction on alterations is automatically amended if the tenant is under one of these duties. If the tenant proposes to make an alteration to comply with the duty, but is not entitled to do so as a matter of his lease or contract with his landlord, the law modifies the restriction to make it easier.

3.58 To qualify, the tenant must have a tenancy (i.e. a lease, sublease or agreement for lease), as opposed to a licence. The tenant must also make an application in writing to begin the process.

For an explanation of the difference between a lease and a licence, see Appendix I. For guidance on making applications for consent, see chapter 5.

3.59 Provided the tenant meets these criteria, the lease is modified so that the tenant can do the alteration with the landlord's written consent, which is not to be withheld unreasonably.

3.60 There are special rules about:

- how long the landlord may take to respond (explained at 6.36 to 6.41); and

- the circumstances in which consent can be withheld and the types of conditions the landlord can reasonably impose (explained at 7.191 to 7.192).

Automatic change 3. Electronic communications equipment

3.61 The law also limits the ability of a landlord to restrict alterations relating to electronic communications services. Merely installing wiring in the demised premises is unlikely to breach a covenant not to make alterations, but this limitation may be useful in the case of more extensive works or more restrictive wording.

3.62 The legislation that introduced this limitation is difficult to interpret, so it is hard to give a clear and straightforward explanation of how it works. In short, certain restrictions are sometimes modified so that the tenant can do the work with the landlord's consent, which is not to be withheld unreasonably.[181]

3.63 The question whether consent has been unreasonably withheld has to be determined having regard to all the circumstances and to the principle that no person should unreasonably be denied access to an electronic communications network or to electronic communications services.[182]

3.64 To see whether the modification applies:

- First, check the *restriction* will be affected.
- Then, check the *purpose* of the tenant's works.

The restrictions

3.65 The law affects provisions that impose any prohibition or restriction on a tenant with respect to an 'electronic communications matter'. Electronic communications matter is defined in s. 134(7) of the *Communications Act* 2003, which is reproduced in Appendix 3. The definition is not straightforward, and further advice should be sought in cases of doubt.

3.66 The provisions must be contained either:

- in a lease, or an agreement for a lease, for a year or more; or
- in an agreement relating to premises to which a lease, or an agreement for a lease, for a year or more applies. This would therefore cover the situation where (for example) the restriction is contained in a side-letter rather than the lease itself.[183]

3.67 In either case, the provisions are only affected if the lease or agreement is entered into on or after 25 July 2003, unless OFCOM orders otherwise.[184] Conversely, OFCOM may also order that particular provisions in leases or agreements after that date should not be affected.[185] The OFCOM website is likely to be a good starting point if you need to find out more.

The purpose of the tenant's works

3.68 The restrictions are modified:

- in relation to things done inside a building occupied by the tenant under the lease; or
- for purposes connected with the provision of an electronic communications service.[186]

3.69 The effect of this provision is unclear. The first limb will certainly cover internal alterations to buildings on the demised premises that relate to an electronic communications matter. It is unlikely to cover any alterations that do not relate to an electronic communications matter, although the wording does not say so. The second limb will also cover alterations to the interior and exterior of buildings and to open land that form part of the demised premises, if the alterations are for the purposes specified. Our view is that those purposes will in practice extend to any electronic communications matter.

3.70 It has been suggested[187] that the first limb also extends to works to common parts that are not let to the tenant but that the tenant has rights over. We disagree: unless the lease contains unusual provisions, no provision imposes restrictions in relation to property that is not part of the demised premises and there are therefore no relevant restrictions to be modified; the tenant is simply granted rights over property retained by the landlord and the section does not extend rights but merely modifies restrictions. It may be possible to overcome problems caused by common parts by using the Electronic Communications Code.[188] Specialist advice should be sought.

Choice of telecoms provider

3.71 The same legislation also affects restrictions on the occupier's choice of providers of electronic communications services. Again, certain restrictions are modified so that the occupier can choose his provider with the landlord's consent, which is not to be withheld unreasonably.[189] Again, the question whether the consent is unreasonably withheld has to be determined having regard to all the circumstances and to the principle that no person should unreasonably be denied access to an electronic communications network or to electronic communications services.[190]

3.72 This modification affects provisions that impose on the occupier a prohibition or restriction under which his choice of:

- the person from whom he obtains electronic communications services; or
- the person through whom he arranges to be provided with electronic communications services

is confined to a person with an interest in the premises, to a person selected by a person with such an interest or to persons who are one or the other.

3.73 The provisions may be contained in a lease, licence or other agreement.

3.74 Again, the provisions are only affected if the lease or agreement is entered into on or after 25 July 2003, unless OFCOM orders otherwise.[191] Conversely, OFCOM may also order that particular provisions in leases or agreements after that date should not be affected.[192]

Discretionary changes

3.75 As explained above, discretionary changes entitle the tenant to apply to court, which may decide to change the restriction.

Discretionary change 1. Improvements to properties let on business tenancies

3.76 Certain business tenants are entitled to carry out improvements regardless of the terms of their lease. The right to do so is contained in Part I of the LTA 1927. This right is not widely known and very rarely used, but can be extremely useful to a tenant. However, tenants must follow the strict procedure set out in the Act. Provided they do so, improvements can be carried out even if the lease expressly forbids them.

3.77 Subject to complying with further procedures, the tenant may also be entitled to compensation from the landlord at the end of the lease. This is explained at 3.88 to 3.92.

3.78 A tenant need only use this procedure if the lease prohibits the intended works. If the works are not restricted, or if the tenant could get the landlord's consent to go ahead, there is no need to do so. However, if the tenant wants the right to claim compensation for the works at the end of the lease, he should use the procedure anyway, and follow it carefully.

Entitlement to carry out improvements

3.79 To use the procedure, the tenant must meet the following criteria:

- The tenant must have a qualifying lease. The following do not qualify:

  mining leases;
  - leases of agricultural holdings (usually agricultural tenancies granted before 1 September 1995);
  - farm business tenancies (usually agricultural tenancies granted on or after 1 September 1995);
  - leases where the premises are 'let to the tenant as holder of any office, appointment or employment from the landlord', so long as the tenant holds that position and provided (if the tenancy was created after 23 March 1928) the tenancy is in writing and expresses the purpose for which it was made.[193]

  A mere licence to occupy will not be sufficient.

For more information on the distinction between leases and licences, see Appendix 1.

- The premises must be used for trade or business.

  The words trade or business include a profession, but exclude the business of subletting the premises to residential occupiers. There is no need for the tenant to be aiming to make a profit. So, for example, a charitable college that retains all income for its own purposes and does not distribute any profit would qualify.[194]

  If the premises are only partly used for trade or business the Act will still apply, but the improvements must relate to the trade or business.

Example

A tenant has a lease of an office, with a residential flat above. Improvements to the ground floor office would qualify, but improvements to the flat above would not, unless the work was somehow 'related' to the trade or business carried on at the office.

- The works must qualify as 'improvements'. There is no definition of improvements, but the tenant will need to prove that the works:

> - are calculated to add to the letting value of the premises at the end of the tenancy;
> - are reasonable and suitable to the character of the premises; and
> - will not diminish the value of any other property belonging to the landlord or any superior landlord.[195]

'Improvements' therefore has a more restricted meaning than it does in the context of transforming qualified covenants (see 3.40 to 3.45).[196] Nevertheless, the complete demolition of a building and replacing of it with another will qualify, so long as it meets these criteria.[197]

In the event of a dispute, both the landlord and the tenant will need evidence to show the impact of the works on the value of the premises or the landlord's other property. The tenant can only proceed if the works have no negative effect on them whatsoever.

3.80 There is some legal uncertainty about whether meeting these criteria alone will entitle the tenant to go ahead. One case indicates that the tenant must also prove he will be entitled to compensation at the end of the lease in order to win the right to do the improvements in the first place.[198] This imposes further restrictions, as explained in 3.88 to 3.92.

However, in our view (and that of other authors[199]) the issue of compensation is separate. Therefore a tenant ought to be able to do the improvements based on the above criteria alone, once the procedure has been followed. Surveyors should be aware, however, that the issue has not yet been resolved, and courts may force tenants to comply with the compensation criteria as well.

The procedure

3.81 To override a restriction in the lease, the tenant must follow the proper procedure.[200] This must be done very carefully to ensure that nothing is missed, and must be completed before work begins. Surveyors should consider taking legal advice. There are five steps to the process.

3.82 *1. Initial notice:* The tenant begins by serving written notice of his intention to make the improvement on the landlord. Chapter 5 explains how to go about serving notices. The tenant must provide a specification of the proposed works and a plan showing how the existing premises are affected (although this does not have to be served with the notice[201]). The landlord will have the option of doing the works himself, so the plan and specification ought to be detailed enough to allow him to do so. It is also good practice to:

- explain that the tenant is making an application under Part I of the LTA 1927, rather than simply an application under the alterations clause in the lease (it is unclear whether this is a strict requirement);[202] and

- enclose evidence that the works would satisfy the criteria for an 'improvement' set out in 3.79.

3.83 *2. Time limit for landlord to object:* Once the notice, plan and specification have been served, the landlord has three months to decide whether to object.

If he does not object, the tenant can proceed regardless of the terms of the lease. It is therefore critical that a landlord should object within the time limit, or he risks losing the chance to prevent the works from going ahead.

The tenant must nevertheless adhere rigidly to the details of the works set out in the initial notice. Only those precise works are permitted and doing anything more would be a breach of the lease. The tenant must also be sure that it served the initial notice correctly, otherwise the landlord would never need to respond.

3.84 *3. Notifying superior landlords:* Even if the landlord does not want to object, he should serve copies of the tenant's initial notice on any superior landlord. The landlord can claim from a superior landlord any compensation he has to pay the tenant – but only if he serves this notice. The time limit for serving the notice on superior landlords is not clear, but landlords would be well advised to act immediately on receipt of the tenant's initial notice.

A tenant, however, only need serve his initial notice on his own immediate landlord.

3.85 *4. Landlord's objection:* If the landlord does object, he can choose one of two options:

- object outright; or
- object to the *tenant* doing the works, but offer to do them himself in return for a reasonable increase in the rent.

If the landlord offers to do the works himself, he can either specify the increase in rent he thinks would be reasonable, or offer to be bound by whatever the court decides. If the landlord specifies the amount, he faces the risk that the court might disagree. The objection might then be invalid and the court would allow the tenant to go ahead.

The tenant has the option of declining the landlord's offer and abandoning the application altogether.[203]

In any case, there is nothing to stop the landlord and tenant from reaching agreement on terms to allow the works to go ahead.

3.86 *5. Application to court:* If the landlord has objected the tenant need not abandon his plans. He can apply to court for an order that the works can go ahead.

The tenant will need to prove that the works meet the criteria set out above (3.79 to 3.80). If the court grants the order, the tenant must still adhere to the description of the works in the initial notice or any modifications made by the court.

3.87 Once the works are done (either because the landlord did not object, or because the court ordered that the works could go ahead) the tenant can ask the landlord to certify that the improvements have been completed. This will help to avoid any dispute when the tenant claims compensation. If the landlord refuses or fails to give the certificate within a month, the tenant can ask the court for an order to the same effect.

Claiming compensation

3.88 If the tenant has complied with the procedure and done improvement works, he is also entitled, in principle, to compensation at the end of the lease.[204] This is subject to the tenant following another detailed procedure and meeting strict criteria.

3.89 The requirements are summarised below, but landlords and tenants should seek legal advice on any claim

3.90 The tenant must show that:

- the lease still met the qualifying requirements, and that the premises were used for trade or business (see 3.79);

- the notice procedure explained in 3.81 to 3.87 was followed before the works began, and the works were completed by the tenant or a previous tenant under the same lease;

- he served a notice claiming compensation within set time limits (which vary depending on the way the lease comes to an end);

- he has 'quit the holding' (i.e. moved out of the premises);

- the improvements qualify for compensation (the Act excludes tenant's or trade fixtures which the tenant is legally entitled to remove, and improvements that the tenant was contractually obliged to make); and

- the improvement adds to the letting value of the premises.

3.91 The amount of compensation is based on the lower of:

- 'the net addition to the value of the holding as a whole which results directly from the improvement'; and

- 'the reasonable cost of carrying out the improvement at the termination of the tenancy' less the cost of any repairs needed to the improvement (unless, as will usually be the case, the tenant is already liable for the disrepair under a repairing clause in the lease).

3.92 The result should be that the landlord pays for any benefit he gets in taking back improved premises at the end of the lease. If the landlord will, in actual fact, not receive any benefit because he (or a new incoming tenant) intends to demolish the premises or use them for a different purpose, the compensation will be reduced.

Checklist for improvements to business tenancies

3.93 Check that there is a restriction and you need to use this procedure – see the checklist at the end of this chapter –

as the tenant may be entitled to do the works anyway. If you do need to use the statutory procedure, or you want to use it in order to claim compensation later, make sure you can answer 'yes' to all the following questions:

- Does the tenant have a qualifying lease?
- Are the premises used wholly or partly for trade or business purposes?
- Do the proposed works meet the criteria of 'improvements'?
- Has the tenant served an initial notice and provided plans and specifications?
- Have three months expired without the landlord serving notice of objection?
- If the landlord has objected, has the tenant obtained a court order that the works can go ahead?

If so, the tenant can go ahead with the works provided he adheres to the plans and specifications served with the initial notice.

Discretionary change 2. Converting dwellings into flats

3.94 Where the conversion of premises into two or more flats would be restricted by a lease, the law gives some limited assistance.[205]

3.95 If the tenant can meet certain criteria, the court can change the restriction to allow the development to go ahead. The criteria are:

- 'owing to changes in the character of the neighbourhood in which the premises are situated, they cannot readily be let as a single dwelling-house but could readily be let for occupation if converted into two or more dwelling-houses'; or
- 'planning permission has been granted under Part III of the *Town and County Planning Act* 1990 for the use of the premises as converted into two or more separate dwelling-houses instead of as a single dwelling-house'.

3.96 These are alternatives, so there is no need for the tenant to satisfy both criteria.

3.97 The tenant must make a court claim to have the lease varied, and the landlord, or any other person interested, is entitled to object. The court has a wide power to change the lease, and can impose any conditions and terms it thinks just.

3.98 The conversion need not involve structural alterations. For example, it was enough where a tenant planned to convert a house into 'one-room flatlets' with shared bathrooms.[206]

3.99 If applying on the first ground (changes in the character of the neighbourhood), the tenant will have to gather evidence of the precise changes he is relying on. The court will also want to see how the changes have *caused* the difficulties in letting the house as a single unit.[207] This is not necessary under the second ground (planning permission), but the court still has the discretion to decide whether to allow the works to go ahead.

Discretionary change 3. Access to neighbouring land

3.100 In some situations the problem for the tenant will not be a restriction in the lease, but the need for access to other land to complete the works. Without a legal right or permission from the owner, a tenant cannot go on to other land for any reason (see 3.4 to 3.6).

3.101 The *Access to Neighbouring Land Act* 1992 gives tenants, and others, the right to obtain a court order giving access. However, this will be of only limited use for a tenant hoping to do alterations. This is because the Act can only be used for 'preservation works' such as maintenance and repair of a building.[208] However, preservation works may include incidental alterations and improvements.[209] The Act may therefore be of some use to a tenant hoping to do alterations as part of a larger programme of repairs.

3.102 The *Party Wall etc Act* 1996 may also give some assistance.

3.103 Specialist advice should be obtained on the operation of the Acts and the relevant procedure.

Discretionary change 4. General

3.104 Tenants of long leases can also apply to the Lands Tribunal for a restriction on alterations to be discharged or modified. This is explained at 4.68 to 4.73.

Other ways to overcome restrictions

3.105 Where works are required by an Act of Parliament, the Act frequently gives the tenant the right to apply to the court to have the lease modified or even ended. In some cases the court can also direct how the costs of the works should be split between the landlord and the tenant or adjust the rent. Examples include the *Fire Precautions Act* 1971, the *Factories Act* 1961 and the *Offices, Shops and Railway Premises Act* 1963. Legal advice should be taken on how any of these Acts operate and the procedure for applying to court.

Summary: Overcoming restrictions

This is a summary only, and reference should be made to the full explanations given above.

	Works covered	Effect	Procedure
AUTOMATIC MODIFICATIONS			
Improvements	Works that are an improvement from tenant's point of view	Qualified covenants become fully qualified	Tenant must apply for consent in usual way
Compliance with disability laws	Any works necessary to comply with tenant's duties	Absolute or qualified covenants become fully qualified	Tenant must apply for consent in usual way
Electronic communications equipment	Works that relate to an electronic communications matter	Absolute or qualified covenants become fully qualified	Tenant must apply for consent in usual way

	Works covered	**Effect**	**Procedure**
DISCRETIONARY MODIFICATIONS			
Improvements to business tenancies	Any works that qualify as 'improvements', which may include total demolition and rebuilding	Improvements can go ahead regardless of restrictions	Tenant must serve notice in advance and apply to court if landlord objects
Converting dwellings into flats	Conversion into flats	Conversion can go ahead regardless of restrictions	Tenant must apply to court
Access to other land	Preservation works and incidental alterations or improvements	Tenant may be granted access to neighbouring land to carry out works	Tenant must apply to court

The effect on rent

3.106 When a tenant carries out alterations it will be important to be aware of the effect they will have on the rent, either at rent review, or when the lease is renewed. Alterations will often improve the value of the premises, yet tenants are likely to feel that this should not be reflected in the rent where the works have been done at their own expense.

Checklist: Carrying out alterations

Use this checklist to decide whether intended works are restricted by the lease, and if so how the restriction operates and whether it can be avoided.

- Are the works within the demised premises? See 3.3. If not, the tenant has no right to do the works.
- Does the tenant need access to other land? See 3.4 to 3.6. If so, see 3.100 to 3.103.
- Do the works require planning permission? If so, does the tenant need the landlord's consent to apply? See 3.23 to 3.24.
- Is there an express restriction in the lease? If so, does it cover the intended works? See 3.10 to 3.20.
- Does the restriction prohibit the works completely or can they be done with landlord's consent? See 1.6 to 1.15.

- Does the lease expressly say that consent is not to be unreasonably withheld? If not, is this implied anyway? See 3.40 to 3.45.
- If there is no express restriction, is a restriction implied? If so, does it cover the intended works? See 3.27 to 3.34.
- If consent cannot be unreasonably withheld, tenants should consider how to make a proper application for consent (see chapter 5), and landlords should consider how to respond (see chapters 6 and 7).
- If the restriction (express or implied) prohibits the works completely, or if consent can be withheld, can the restriction be avoided? See 3.52 to 3.105.

4

Changing use

This chapter explains:

- the tenant's position if there are no restrictions;
- the meaning of common lease clauses;
- when the landlord is prevented from demanding a premium;
- when the landlord must act reasonably; and
- how to overcome restrictions.

General principles

The starting point: the tenant is free to use the premises for any purpose or none

4.1   The tenant is free to use leasehold premises for any purpose, unless there is a restriction. Most restrictions relating to use are express (written down in the lease, usually as a tenant's covenant).

4.2   Unless there is a positive obligation to use the premises (see 4.9 to 4.12), a tenant is free not to use the premises at all, even if the lease prohibits all uses except one named use.[210]

4.3   If the tenant fails to comply with an obligation or breaches a restriction, the landlord will be entitled to use the remedies explained in chapter 10.

General restrictions

4.4   The tenant's right to use the premises for a particular purpose may also be restricted by the general law, most notably by planning legislation. There is usually no implied warranty on the part of the landlord that the property can be used lawfully for any specific purpose.[211] The tenant should therefore check the planning position even if a particular use is expressly permitted by the landlord.

4.5 A further notable statutory restriction is against use as a brothel. It is an offence for the tenant or occupier, or person in charge, of any premises knowingly to permit the whole or part of the premises to be used as a brothel. If the tenant is convicted, the landlord may require the tenant to assign the lease to some person approved by the landlord, such consent not to be unreasonably withheld. If the tenant fails to do so within three months, the landlord may terminate the lease. If the landlord does not exercise his rights, or if he grants a new tenancy to the tenant, and the tenant re-offends, the landlord risks prosecution as well.[212]

4.6 There may also be specific restrictions affecting the landlord's own ability to use the premises. If the landlord owns the freehold, it may be subject to a freehold restrictive covenant. If the landlord has a lease himself, there may be restrictions in the lease. These restrictions are likely to be enforceable against a subtenant and any other occupier of the land.[213]

Types of use covenant

4.7 Restrictions or requirements in the lease are usually found in the tenant's covenants. Frequently the tenant will covenant not to use the premises other than for the 'Permitted Use', which may be defined in a list of 'Particulars' or definitions at the beginning of the lease.

4.8 Use covenants can be positive or negative.

Positive covenants

4.9 A positive covenant requires the tenant to make some use of the premises, usually a specified use. The tenant will therefore be in breach of the covenant if he does not use the premises at all.

4.10 Positive covenants in leases of retail shops often require a tenant to trade during normal business hours. For many businesses, 'normal business hours' now include Sundays. But if the lease was entered into before 26 August 1994 and has not been varied after that date, a retail tenant is not normally required to open his shop on Sundays.[214] The main exception to this rule is where the lease specifically relates to Sundays and requires Sunday trading of a kind which would have been permitted under the previous legislation.[215]

4.11   A covenant that appears to put a positive obligation on the tenant may not actually have that effect; the courts have sometimes interpreted an apparently positive covenant as an emphatic negative covenant. In those cases the covenant is deemed to prevent the tenant from using the premises for anything other than a specified use, but not positively to require him to use the premises for the specified purpose.

Example

A tenant covenanted '*not to carry on upon the demised premises [certain named offensive businesses] or permit to be carried on in the premises any other trade business or calling whatsoever that shall constitute a nuisance or annoyance to any other tenant or tenants of the Lessors but will use the demised premises either for the business of high-class retailers of jewellery and/or antiques and/or luxury goods and/or a travel agency or a recognised Bank the authorised name of which includes the word "Bank"*'.

Despite the positive language of the second part of the clause, the judge held that this was simply an emphatic statement of the prohibition on uses other than those specified.[216]

4.12   Even if there is a positive obligation to use the premises, it may be sufficient if the tenant only uses part of the premises for the specified purpose and leaves the rest unused.[217]

Negative covenants

4.13   Negative covenants simply restrict the use that can be made of the premises. An example of a negative covenant is '*not to use the premises otherwise than as barristers' chambers*'.

Express restrictions

4.14 Restrictions can be absolute (meaning a complete prohibition on one or more types of use) or qualified (meaning that the landlord's consent is required to use the premises for one or more types of use). Many qualified covenants are fully qualified (meaning that they provide that consent cannot be withheld unreasonably). Covenants can also be hybrid: if a condition is not satisfied, the covenant is of one type (for example, absolute); if it is satisfied, the covenant is of another type (for example, fully qualified).

> For more information on the different types of covenant, and the importance of distinguishing between them, see 1.6 to 1.17.

4.15 Unlike alienation and alterations covenants, qualified use covenants are *not* made fully qualified by law. However, the law does usually prohibit the landlord from demanding a premium if the change of use does not involve structural alterations: see 4.53 to 4.60.

Understanding the restriction

4.16 Express restrictions come in various forms. It is crucial to be able to understand precisely what they mean, because the premises may be used for anything not covered by a specific restriction without consent (provided the use is not restricted in any other way). Each covenant should be examined separately to see if the proposed use is restricted.

4.17 Interpreting words in leases is a difficult task and it is often wise to take advice. Previous court decisions can be used as a guide to interpretation, but only where the words of the lease and all the circumstances are the same.

4.18 However, some general guidance on how to interpret covenants is given at 1.65 to 1.71 and more specific guidance for use covenants is set out here. In addition, there are some key terms that appear in large numbers of leases, the meaning of which is relatively certain. These are explained at 4.23 to 4.44.

Ambiguous wording

4.19 Where the wording is ambiguous, the covenant is likely to be interpreted as permitting the tenant to do more rather than less.[218] This is the case even where it is against the tenant's interest (for example, where the tenant wishes the use clause to be more restricted in the hope that this will reduce the rent determined on a rent review). However, as with most general rules, this principle will only apply where it is not otherwise possible to choose between two possible meanings.

Physical nature of the property

4.20 The wording should be interpreted in the light of the physical nature of the property at the time the lease was granted.

> *Example*
>
> A tenant took a lease of a shop with living accommodation above. He covenanted to '*use and occupy the demised premises ... for the trades or business of a newsvendor stationer bookseller toy merchant and tobacconist only*'. The court decided that he was entitled to live in the residential accommodation without being in breach of covenant. In view of the nature of the property, the covenant was interpreted as applying only to the business portion.[219]

Changes in the meaning of words

4.21 Words may change their meaning over time. If they have done so, then the meaning at the date the covenant was entered into will apply. The meaning will be determined in the context of any other restrictions and set against the circumstances in which the lease was granted.[220]

> *Examples*
>
> (1) A lease granted in 1950 contained a covenant against carrying on the business of a '*coffee house keeper*'. The words could be traced back to restrictions in a standard estate lease from the 1790s. At that time, they would have been understood to refer to a seventeenth or eighteenth century coffee house (more like an inn). Instead, the court held that they referred to a place where coffee

and light refreshments were served, which was how the words would have been understood in 1950.[221]

(2) A lease granted in 1947 restricting use to that of a hairdresser also permitted the sale of contraceptives.[222] But a lease with the same use covenant granted after contraceptives became more readily available would probably be interpreted more restrictively.

4.22 Sometimes the lease will expressly provide that the meaning of the covenant may change in the future. This is most commonly found in leases that describe the permitted use by reference to planning legislation and expressly state that subsequent changes to the legislation are to be taken into account.

Example

If the lease contains a covenant '*not to use the premises except for a purpose falling within Class A1 of Schedule 1 to the Town and Country Planning (Use Classes) Order 1987 as amended or re-enacted from time to time*' then if the text of the Order is amended after the lease is entered into, the new wording rather than the original wording will apply.

The whole of the lease should be checked. Sometimes there will be wording about amendments and re-enactments in the interpretation section of the lease, rather than in the covenant itself. Care must be taken: for example, does the interpretation section say only that amendments and re-enactments of Acts are to be taken into account? If so, amendments and re-enactments of statutory instruments are likely to be irrelevant.[223]

If the wording of the covenant is simply '*not to use the premises except for a purpose falling within Class III of The Use Classes Order 1972*', subsequent changes in the legislation will usually be irrelevant. The only uses permitted under the lease will be those which were included in Class III at the date the lease was entered into.[224]

The meaning of common lease clauses: general covenants

4.23 Leases frequently contain general covenants relating to use, in addition to or instead of specific covenants.

Examples

(1) A lease contains:

- a specific covenant not to use the premises other than as a nightclub; and
- a general covenant not to use the premises so as to cause a nuisance or annoyance to the landlord or to the landlord's other tenants.

The tenant will have to operate the nightclub in such a way as to avoid causing a nuisance, for example by making sure the sound-proofing is adequate. The covenant against nuisance will be interpreted in such a way as to make it possible to operate a nightclub.

(2) A lease contains no specific covenants relating to use, but does contain a general covenant not to cause a nuisance or annoyance to the landlord or to the landlord's other tenants. The tenant will not need permission to change use, but he will only be able to use the premises for a use which does not cause a nuisance or annoyance.

'Nuisance'

4.24   Nuisance bears a strict legal meaning. On the present state of the law, covenants not to cause a nuisance refer to nuisance in the legal sense,[225] although this has been doubted.

'Annoyance'

4.25   Covenants that restrict annoyance prevent more activities than covenants that merely restrict nuisance. Anything that reasonably disturbs the 'peace of mind' of an 'ordinary sensible' person will be an annoyance, whether or not it amounts to a physical detriment to comfort.[226]

'Immoral purposes'

4.26   Use for prostitution is use for an immoral purpose.[227]

4.27   In the early twentieth century, use of premises by a kept woman, whose rent was paid by her lover who visited regularly, was use for immoral purposes.[228] It is extremely unlikely that a covenant against immoral use would be interpreted the same way in a modern lease.[229]

4.28   It is unclear whether a covenant against immoral use would refer to use considered immoral at the date of the lease or at the date of the breach. If the date of the lease is the correct date, a covenant in a lease granted before the 1960s might be construed as prohibiting use by unmarried couples. It might also be construed as prohibiting use by civil partners, but any attempt to bring proceedings against civil partners would almost certainly be prevented by anti-discrimination legislation[230] and/or the *Human Rights Act* 1998. However, we consider that the correct date is the date of the breach.

'Offensive trades'

4.29   A covenant against offensive trades does not impose a blanket prohibition on any specific trades. Instead this restriction focuses on whether the trade carried on is offensive in the actual circumstances of the particular case. So a trade may be offensive if it is carried on badly, but may not be offensive if it is carried on properly. Similarly a trade may be offensive in one location, but not in another.[231]

Use 'for the purposes of' a business

4.30   Premises can be used for the purposes of a business even if the main business activity is carried on somewhere else.

> *Example*
>
> If the premises are used to store trading stock or for administrative support for a retail shop on another site, this is likely to be use 'for the purposes' of that retail business.[232]

Use for a named business

4.31   A prohibition on carrying out a named business will not prohibit every act which forms part of that business. There will only be a breach of the covenant if the level of activity actually amounts to the carrying on of the prohibited business.

Example

If the lease contains a covenant 'not to carry on the trade or business
of selling bread and confectionery', a tenant with a grocery business
could sell small quantities of bread and confectionery.

But if the lease contains a covenant 'not to sell bread and
confectionery by way of trade or business', the tenant could not sell
any bread or confectionery as part of his business.[233]

4.32 Sometimes a tenant can fairly be said to be carrying on
two or more businesses at the same time. If one of those
businesses is a prohibited business, the tenant will be in
breach of the lease.

Example

If the lease contains a covenant not to use the premises other than
as a supermarket, the tenant will not necessarily be in breach of the
covenant if he sells some sweets and cigarettes as well. But if he
sells sufficient quantities of tobacco and confectionery from a large
display area, giving a high percentage turnover from those products,
he may be carrying on the trade of tobacconist and confectioner as
well as the business of a supermarket, and will be in breach of
covenant.[234]

'Trade' and 'business'

4.33 Prohibitions on trade are more limited than prohibitions
on business. 'Trade' refers only to businesses conducted
by buying and selling.[235] 'Business' refers to anything
that is an 'occupation rather than a pleasure'.[236]

4.34 Covenants against carrying on business are often found
in residential leases. Whether an activity is a business or
not will depend on whether there is direct commercial
involvement and whether the activity is more than
ancillary or subordinate to the residential use.

Example

A tenant used a flat for regular meetings of a political and social
committee and for the administration of the committee. The

landlord accused him of breaching a covenant '*not to carry out any profession, trade or business on the premises*'.

The judge found that the tenant's activities in his flat connected with the committee constituted an occupation or duty as distinct from a pleasure or a leisure activity, and that his work for the committee was to him 'a serious undertaking earnestly pursued for the purpose of fulfilling a duty assumed by him'. His work for the committee was something quite diverse from, and not merely incidental to, his ordinary domestic life; it was continuous and regular and not merely sporadic. It went beyond activities that were merely ancillary to normal domestic life, and so the tenant was in breach.[237]

4.35 Taking in lodgers and subletting of multiple rooms are both likely to be a breach of a covenant not to carry on a business.[238] A single lodger or single subletting is less likely to be considered a business.

4.36 A covenant not to permit business to be carried on on any part of the demised premises is broken by letting the gable-ends of a house to a bill-posting firm.[239]

'A'

4.37 Use covenants sometimes prohibit the use of premises otherwise than as 'a' private dwelling-house or other named use. Depending on the context, this can mean 'a single' private dwelling-house,[240] or it can refer only to the type of use and not the number of units.[241] It is less likely to permit the subdivision of a single house into flats than it is to permit the use of several houses each as a private dwelling-house. The question is therefore often closely related to alterations.

The meaning of common lease clauses: specific covenants

4.38 Any specific covenant (i.e. one that prescribes or prohibits a specific use) may be inserted in a lease. Some common covenants have been considered by the courts, who have given guidance as to what they mean. Only a few are covered here.

'Private dwelling-house'

4.39 A covenant not to use premises other than as a private dwelling-house prohibits:

- conversion of a house into flats[242] and multiple sublettings of a house in parts;[243]
- receiving paying guests on a regular basis;[244]
- using the premises for the business of granting licences for occupation;[245]
- using the premises for a private car hire business by receiving telephone calls at the house and keeping a car in the garage, even though no customers call at the house;[246] and
- using the premises for holiday lets.[247]

4.40 It does not prohibit:

- use by a group of students who live communally;[248] or
- a sale by auction of the house's furniture.[249]

4.41 Whether or not it prohibits a single subletting of part of the house is particularly dependent on the context. For example, if another clause in the lease contemplates subletting in parts, a single subletting is less likely to be a breach of this covenant.[250]

'Supermarket' and similar words

4.42 The developments in supermarket trading over the course of the twentieth century mean that great care is needed when interpreting covenants involving this type of use.

Examples

(1) A 1950s covenant restricted the tenant to using the premises for the business of '*grocers provisions wine spirit and beer merchants*'. This did not permit the sale of fresh fruit and vegetables, fresh meat or fish, or newspapers and magazines. The sale of newspapers, magazines and stationery, records, tapes, videos for hire and electrical goods in addition to grocery goods and other food meant that the tenant was either carrying out several trades, including grocer, newsagent and hirer of video films, or was trading as a general store.[251]

(2) A 1960s covenant restricting the premises to use as a supermarket permitted the tenant to sell freezer cabinets generating 15 per cent of the shop's turnover.[252]

(3) A 1980 covenant restricted the tenant to using the premises effectively as a food supermarket. This allowed the tenant to sell household articles customarily bought in a small food supermarket (which were not defined, but would include some toiletries, cleaning products and so on). But it did not permit the sale of items such as babies' bottles, scissors or electrical plugs and batteries.[253]

(4) A late twentieth or twenty-first century lease granted for use as a supermarket (as opposed to a food supermarket or any other specialised form of supermarket such as a DIY supermarket) is likely to permit the sale of a very wide range of goods.

'Car repairs'

4.43 A covenant permitting car repairs will permit bodywork and spraying as well as mechanical repairs.[254]

4.44 A covenant permitting use for the business of 'a garage with car sales and vehicle repairs' will allow the tenant to sell some spare parts, such as fan belts, and items that might be fitted by a garage, such as child seats and roof racks. It will not allow the tenant to sell oil and lubricants, or general motoring accessories such as first aid kits, torches and car cleaning aids.[255]

Impact of other covenants

4.45 Other covenants can sometimes form an indirect restriction on use.

Compliance with statutes

4.46 Most well-drafted leases require tenants to comply with any statutes or regulations affecting the use of the premises. This will sometimes impose an indirect restriction on uses that would otherwise be permitted (for example, a use that would put the property in breach of fire regulations).

Not to apply for planning permission without consent

4.47 Many leases restrict the tenant's right to apply for planning permission for a change of use without the landlord's consent. This presents another potential restriction for the tenant if the change would require planning approval. Although the lease may not restrict the use as such, if the tenant cannot implement the change without planning consent, he must get the landlord's approval before applying

4.48 The landlord need not act reasonably, unless there is express wording in the lease requiring him to do so.[256] He may also demand a premium in return for his consent. Such a clause may therefore enable a landlord to prevent a change of use unreasonably, even if his consent to the use itself may not be unreasonably withheld.

Example

A lease contains a fully qualified covenant *'not without the landlord's prior written consent (such consent not to be unreasonably withheld) to use the demised premises otherwise than as a hairdresser'*. The lease also contains a qualified covenant 'not to make any application for planning permission without the prior written consent of the Landlord'.

If the tenant wishes to use the premises as a hot food takeaway within Use Class A5 and the premises only have planning permission for use within Class A1, then looking only at the first covenant, it seems that the landlord can only refuse consent if it is reasonable to do so. But the second covenant will allow him to refuse consent for the tenant to apply for planning permission, and he need not act reasonably in doing so. If the tenant applies for planning permission without consent, he will be in breach of the lease.

Implied restrictions

4.49 While it is not impossible for a use restriction to be implied, this will be extremely unusual. Legal advice should be sought if the question arises.

Additional tenant protection

4.50 In some circumstances the law gives the tenant additional protection. This comes in two forms:

- Sometimes the law prevents the landlord from demanding a premium.

- Sometimes the law imposes a requirement for the landlord to act reasonably.

The position depends on whether the wording of the restriction is absolute, qualified or fully qualified.

For an explanation of absolute, qualified and fully qualified covenants, see 1.6 to 1.15. For more information on what constitutes a premium, see 8.19 to 8.22.

Absolute restrictions

Premiums

4.51 In cases where the use is absolutely prohibited, if the tenant asks for consent the landlord is entitled to ask for a premium as the price of his consent.[257]

Acting reasonably

4.52 Absolute covenants are not made fully qualified. They always constitute a complete prohibition unless they can be overcome in one of the ways described at 1.9.

Qualified restrictions

Premiums 1: no structural alterations

4.53 If there will be no structural alterations, then in most cases s. 19 of the LTA 1927 will prohibit the landlord from asking for a premium.[258]

See Appendix 2 for details of when this prohibition does not apply.

4.54 Where this rule applies, it is impossible to 'contract out'. Therefore no attempt in the lease to avoid it (whether deliberate or not) will have any effect. However, there is an important exception: the provisions might be interpreted as setting out conditions which must be satisfied before the tenant can make an application for consent; see 1.36 to 1.40. In addition, some commentators suggest that there is nothing to prevent a landlord from refusing consent, but offering to enter into a new lease at a higher rent, or for a premium, with a different use clause.[259]

4.55 However, the landlord may be entitled to compensation. If the change of use will damage the premises or any neighbouring premises owned by the landlord or reduce their value, he is entitled to demand payment of reasonable compensation as a condition of giving consent.[260] If there is an RTM company (see 5.23 to 5.25), the RTM company has no right to compensation.[261]

4.56 In contrast to requests for consent to make improvements, even if loss of value is the landlord's only objection to the change of use, consent can be refused. The landlord is only prohibited from demanding a premium; he is not prohibited from refusing consent unreasonably.

4.57 The way in which the landlord expresses his request for compensation is important:

- If the landlord said, 'I give my consent provided you pay me £x (an unreasonable amount) and I will not take a penny less', he would probably be found to have demanded a premium and the tenant will be free to go ahead.
- If the landlord said, 'I give my consent provided you pay a reasonable sum as compensation, and I am advised and consider that £x is such a sum',[262] the amount should be determined by the court and the sum paid by the tenant in return for the giving of consent.[263]

It is sensible for the landlord to put forward an amount that he considers reasonable, but state that he will accept such other sum as may be found to be reasonable by the court.

4.58 The landlord (or RTM company – see 5.23 to 5.25) is also entitled to ask for payment of 'any legal or other expenses properly incurred in connection with such licence or consent'.[264] The types of costs the landlord can recover are dealt with in chapter 8. Any dispute as to the reasonableness of the costs demanded can also be determined by the court.

Premiums 2: structural alterations

4.59 If structural alterations are involved, then the landlord will be entitled to demand a premium. 'Involves' simply means that the alteration is part and parcel of the proposal for the change of use; it need not be a necessary element of the change.[265]

4.60 The landlord must, however, take care to state that the premium is for the change of use and not for the alterations. This is because in most cases a structural alteration will qualify as an 'improvement', meaning that a qualified covenant will become fully qualified: consent to the alteration may not be withheld unreasonably.

See 3.40 to 3.44 for more information on improvements.

Acting reasonably

4.61 Qualified use covenants are *not* made fully qualified by law. Nor is this usually implied by any other method.[266] In certain circumstances, there may be legal arguments for implying a proviso that consent is not to be unreasonably withheld[267] (for example if a particular use is contemplated by the covenant but subject to approval of details), but these circumstances are likely to be rare. This may also apply to other situations, such as alterations which do not qualify as improvements.

Fully qualified restrictions

Premiums

4.62 In cases where the landlord may not withhold consent to a change of use unreasonably, the landlord should not ask for a premium for giving consent. In such cases, the landlord is likely to be regarded as having withheld consent unreasonably, so that the tenant can go ahead with the works.[268]

4.63 In unusual cases the lease might specify that the landlord may demand a premium. If the improvement does not involve structural alterations, this provision will be void unless it is interpreted as setting out conditions which must be satisfied before the tenant can make an application for consent: see 4.53 to 4.54. If it does involve structural alterations, we consider that the landlord may demand a premium, provided he limits himself to what is permitted by the lease.

Acting reasonably

4.64 In the case of fully qualified restrictions, the landlord is by definition required to act reasonably.

Overcoming restrictions

4.65 The opportunities for overcoming restrictions on use are limited.

Occupation by persons with mental disorders

4.66 Any agreement that has the effect of prohibiting or imposing any restriction on:

- occupation of a dwelling by persons with mental disorders; or

- the provision of accommodation within a dwelling for such persons

is void to the extent that it has that effect.

This applies automatically provided the property is or includes a dwelling, and the lease or agreement was made after 1 November 1993.[269]

Converting dwellings into flats

4.67 Where the conversion of premises into two or more flats would be restricted by a use covenant in the lease, the tenant can apply to the court to have the lease varied. This modification is discretionary and is explained at 3.94 to 3.99.

Application to the Lands Tribunal for modification or discharge

4.68 Some tenants can also apply to the Lands Tribunal to have a use (or alterations) covenant modified or discharged.[270] The covenant must be a negative one; the section does not apply to positive covenants.[271] To make an application, the following requirements must be satisfied:

- The lease was granted for a term of more than 40 years.[272]

The lease must last for at least 40 years from the later of the date of grant and the date of commencement;[273] so a lease granted on 1 January 1980 for a term of 45 years from 1 January 1980 would qualify, but a lease granted on 1 January 1990 for a term of 45 years from 1 January 1980 would not.

- At least 25 years have already expired.[274]
  The 25 years must also be reckoned from the later of the date of grant and the date of commencement.[275]

- The lease is not a mining lease.[276]

There are further restrictions[277] but these will rarely be relevant.

4.69 If the lease complies with these requirements, then the tenant may apply to vary the restriction. This applies even if it is contained in a variation to the lease made less than 25 years before the application (although in these cases the tenant will succeed very rarely[278]).

4.70 The Lands Tribunal may discharge or modify the restriction if it is satisfied that:

- 'by reason of changes in the character of the property or the neighbourhood or other circumstances … which the Lands Tribunal may deem material, the restriction ought to be deemed obsolete'; or

- 'the continued existence [of the restriction] would impede some reasonable user of the land' and either it:
  - 'does not secure to persons entitled to the benefit of it any practical benefits of substantial value or advantage to them'; or
  - 'is contrary to the public interest'

  and (in either case) 'money will be an adequate compensation for [any] loss or disadvantage which any such person will suffer' as a result of the discharge or modification; or

- all persons entitled to the benefit of the restriction have expressly or impliedly agreed to it being discharged or modified; or

- the discharge or modification of the restriction will not injure the persons entitled to the benefit of it.[279]

4.71 If the discharge or modification is granted, the tenant may have to pay compensation to the landlord.[280]

4.72 If the tenant agrees, the Lands Tribunal may impose further restrictions to take account of the relaxation of the existing ones. If the tenant does not agree, the Lands Tribunal may refuse to make the discharge or modification.[281]

4.73 The tenant can make an application even after he has changed the use and the landlord has brought enforcement proceedings. If the landlord has applied for an injunction or damages, the tenant can apply to have the landlord's proceedings stayed pending determination of his application.[282] He cannot get a forfeiture claim stayed,[283] but he can still apply to the Lands Tribunal after the forfeiture.[284]

Competition law

4.74 Some restrictions on use may be affected by laws on trade and competition. These are outside the scope of this book.

The effect on rent

4.75 When the permitted use is changed it will be important to be aware of the effect this will have on the rent when the lease is renewed or (depending on the terms of the review clause) at rent review. A different use may improve or diminish the value of the premises.

Checklist: Changing use

Use this checklist to decide whether a proposed use is restricted by the lease, and if so how the restriction operates and whether it can be avoided.

- Are there any restrictions affecting the landlord that also affect the tenant? See 1.61.
- Does the change of use require planning permission? If so, does the tenant need the landlord's consent to apply? See 4.47 to 4.48.

- Is there an express restriction in the lease? If so, does it cover the intended use? See 4.16 to 4.44.
- Does the restriction prohibit the change of use completely or can it be changed with landlord's consent? See 1.6 to 1.15.
- If the use can be changed with landlord's consent, does the change of use involve structural alterations? See 4.59. If not, the landlord cannot demand a premium.
- Does the lease expressly say that consent is not to be unreasonably withheld?
- If consent cannot be unreasonably withheld, tenants should consider how to make a proper application for consent (see chapter 5), and landlords should consider how to respond (see chapters 6 and 7).
- If the restriction prohibits the change of use completely, or if consent can be withheld, can the restriction be avoided? See 4.65 to 4.73.

5

Applying for consent

This chapter explains:

- the preliminary checks tenants should make before applying;
- how to ask for consent;
- what an application should include (with sample letters); and
- how to 'serve' an application.

Before you begin: Preliminary checks for tenants

5.1 Many applications for consent are made when there is no need for them (because the tenant is free to go ahead anyway), or when the landlord is not obliged to respond (because the intended action is completely prohibited).

5.2 If a tenant applies for consent when there is no need, or when the landlord is not obliged to respond, he may waste time and incur unnecessary expense. He may also agree to restrictions on the intended action if he mistakenly believes that he needs the landlord's consent.

5.3 To avoid these potential problems, tenants should check the restrictions in the lease before they make an application for consent. The following checklist explains how to do this.

Is an application appropriate?

5.4 Before you begin:

- Check the lease and any ancillary documents, such as a superior lease or a previous licence, to find out whether the intended action is expressly restricted.
- Check the relevant section of this book (2.36 to 2.78 on alienation, 3.27 to 3.34 on alterations and 4.49 on change of use) to see if a restriction will be implied.
- If there is a restriction, does it cover the intended action? For example, there may be a restriction on subletting *part* of the demised premises, but is the tenant free to sublet the *whole*? Similarly, there may be a restriction on structural alterations, but do the intended works actually constitute 'structural alterations'?
- Does the restriction prohibit the intended action completely, or can it be done with the landlord's consent? See 1.6 to 1.15.
- Does the lease expressly say that consent is not to be unreasonably withheld (i.e. is it fully qualified)? If not, check the table at 1.19 to see if this will be implied.

5.5 Based on those checks, the tenant can understand his position:

- If there is no express, or implied, restriction that covers the intended action, the tenant may go ahead without consent.
- If there is a restriction that completely prohibits the action, the tenant may ask the landlord to consent, but this will not be governed by the lease. The landlord may refuse consent for any reason and need not even respond. The tenant should, however, check the possibility of overcoming the restriction. See 2.92 to 2.100, 3.52 to 3.105 or 4.65 to 4.73 depending on the type of covenant involved.
- If there is a qualified restriction that the law does not treat as fully qualified (i.e. qualified covenants against changing use and limited other covenants), the position is the same as a complete prohibition.

However, in most cases other than change of use involving structural alterations, the landlord may not demand a premium

> See Appendix 2 for details of which covenants are not made fully qualified. For details of when a premium is prohibited, see 2.79 to 2.91, 3.35 to 3.51 and 4.50 to 4.64.

- If there is a fully qualified restriction (i.e. landlord's consent must not be unreasonably withheld), the tenant may apply for consent and the landlord must then act reasonably. The tenant should, however, check that any valid pre-conditions can be met.

> For information about pre-conditions, see 1.36 to 1.40.

5.6 Even where a landlord is not obliged to respond, it may still be in the tenant's interest to ask for consent. The landlord may inadvertently give consent (see 9.15 to 9.32), or may simply be willing to do so. This is especially so if the intended action might be to the landlord's benefit (for example, if the tenant wants to assign the lease to a blue chip company or improve the premises).

Making the application

5.7 If the tenant requires consent, it is essential that a proper application is made. This is the trigger that begins the process of obtaining consent, without which the landlord will be under no obligation to respond.[285]

5.8 To make an effective application, it is important that the letter:

- does *not* say 'subject to contract', 'subject to licence' or 'without prejudice';
- does *not* ask for consent 'in principle'; and
- does *not* refer to 'draft' proposals.

In any of these cases the landlord might reasonably conclude that no formal application has yet been made.

If the tenant's proposals may change, it is likely to be in his interest to make an application based on the current intentions and submit a revised application later if necessary.

5.9 The following guidance applies to requests for consent *under the lease*. If the tenant is using a special procedure to overcome a restriction in the lease (for example, the procedure for improvements to properties let on business tenancies explained at 3.76 to 3.93) there may be special rules about how to apply. Always follow the specific guidance that applies in special cases.

The application should be in writing

5.10 The tenant should always apply in writing. This is good practice in every case, since a document will provide evidence of many points that may become important later: the date the application was made, where it was sent, precisely what was applied for, and what was enclosed.

5.11 If the tenant is applying to assign or sublet, there is an additional reason to apply in writing: a written application will attract the protection of the LTA 1988 but an oral one will not. This Act places the landlord under a duty to make a decision within a reasonable time, and not to refuse consent unless it is reasonable to do so. The tenant can sue the landlord for compensation if he fails to comply. This is, therefore, a very important part of the tenant's armoury and tenants will want to ensure it is effective.

> For more information on when the LTA 1988 applies and what the landlord's duties are, see 1.22 to 1.31.

5.12 There are sample application letters at the end of this chapter.

The application should be sent from, or on behalf of, the tenant

5.13 Unless the lease contains unusual terms, the tenant is the only person with the right to ask for consent. The application must therefore be made by the tenant.

117

5.14 There is nothing to stop the tenant's surveyor or solicitor (or anyone else) making the application on the tenant's behalf, provided that is made clear.

Applications made by intended assignees or subtenants

5.15 Many applications are made in connection with an intended assignment or subletting. In those situations, it is not uncommon for the intended assignee or subtenant to make the application. This might be convenient because the intended assignee will have the best information about his finances, easier access to references, and details of any intended alterations or change of use. Nevertheless, to avoid any arguments about whether the application is valid, it should be made by the tenant (or on his behalf).

Example

A tenant intends to assign its lease. The premises are currently used as a shop. The intended assignee will want to change the use of the premises to a fast food outlet and do appropriate alterations.

It might be convenient for the assignee to make the application for consent, because it will need to supply copies of its accounts, references, plans and specifications for the works and details of its intended use of the property. However, the better approach will be for the tenant to obtain that information from the assignee in advance, and to make the application itself.

5.16 Where an application is made by someone other than the tenant, whether it is valid is likely to depend on precisely what is contained in the application, how much the landlord knows about what is going on, and all the other background circumstances.[286]

The identity of the 'tenant'

5.17 The person who is the 'tenant' is the person in whom the legal title to the lease is vested.[287] This may be different from the person who trades from or occupies the property.

5.18 If the lease is registered at the Land Registry, checking the tenant's identity is a simple matter of obtaining copies of the Land Registry entries for the property. The transfer of a registered lease is not complete for these purposes until the assignee is registered at the Land Registry.[288]

5.19 If the lease is not registered, there should be a careful check of any previous assignments to be sure of the identity of the current tenant.

The application should be addressed to the landlord

5.20 Similarly, unless the lease contains unusual terms, all applications should be addressed to the landlord.

The identity of the 'landlord'

5.21 If the landlord's interest in the property is registered, his details can be obtained from the Land Registry in the same way as the tenant's.

5.22 If not, it may be quite difficult for the tenant to ascertain the landlord's identity, depending on the circumstances. Demands for rent or other charges payable under a residential tenancy ought to contain the landlord's name and address.[289] These details are also frequently given with demands for rent for non-residential property.

RTM companies

5.23 Leaseholders of flats have the right to take over management of their building, subject to meeting qualifying criteria.[290] The right is exercised through a special company established for the purpose, known as an RTM (Right to Manage) company. There are special rules about how RTM companies must deal with applications for consent[291] and how they may serve notices.[292]

5.24 The RTM company is deemed to take over the functions of a landlord under a long lease relating to the grant of 'approvals'. The phrase 'long lease' is defined in the governing Act, and includes leases granted for more than 21 years and certain other specific categories of lease.[293] 'Approvals' includes licences and consents.[294]

5.25 In most such cases, the application for consent should be addressed to, and served on, the RTM company.[295] As explained at 6.44 to 6.47, the RTM company has a duty to give notice to the landlord before giving consent. It would therefore probably help speed up the process for the tenant to send a copy of the application direct to the landlord as well.

However, if the premises to which the application relates are not held under a lease by a 'qualifying tenant'[296] (for example, a ground floor shop), it seems the application should be served on the landlord.

Superior landlords and management companies

5.26 Leases occasionally require that the tenant obtains a superior landlord's consent as well as that of the immediate landlord. Tenants should therefore check the lease and, if necessary, make a parallel application to the superior landlord direct. The same might apply to a management company.

5.27 Even if there is no such obligation, the tenant may know that the landlord will need to obtain the superior landlord's consent under the terms of the superior lease. If that is the case, it will help speed up the process to send a copy of the application direct to the superior landlord at the outset.

See 1.53 to 1.54 on when someone else's consent is needed.

Explaining the application

5.28 The tenant should give sufficient detail of the nature of the application for the landlord to understand what he is being asked to consent to and make an informed decision.

5.29 This should include a general description of the transaction, alterations or change of use and refer to any further information the tenant is providing (see 5.32 to 5.52). The tenant should, at least:

- explain which part of the property is affected (or specify that it is the whole);

- in the case of an alienation covenant, identify the other party to the transaction (for example, the intended assignee or subtenant);

- in the case of alterations, identify the general nature of the works;

- in the case of change of use, specify the nature of the new use of the property;

- specify the clause in the lease under which consent is being sought (unless the tenant is applying for consent to carry out a prohibited action, in which case the landlord's attention should not be drawn to the clause in the lease); and

- if there have been previous informal exchanges with the landlord, make it clear that it is intended to put the application on a formal footing.[297]

5.30 This is not all that the tenant ought (as a matter of good practice) to include, but in our opinion it should technically be enough to begin the formal application process. At least in the case of assignment and subletting, it is for the landlord to ask for any additional information he requires.[298]

5.31 The tenant must be careful to ensure that the contents of the application are accurate. If the landlord gives consent (or withholds it in a way that allows the tenant to go ahead anyway), the tenant may only do *exactly* what was asked in the application. If, for example, the tenant's intended alterations have changed, or he wants to assign to a different person, he will need to re-apply.

What to include

5.32 If the tenant intends the application to carry additional weight, and to speed up the application process, there is a great deal more information that should be included. Ideally, this information will be provided at the outset. However, the application will not be invalidated if it indicates that further information (references, for example) will be provided later, so long as it is clearly intended to be treated as a formal application.[299]

5.33 In addition to the basic information (see 5.29), consider providing the following additional details, according to the type of application.

Assignments and sublettings

5.34 The main aims for the tenant will be:

- to demonstrate that any pre-conditions will be met;
- to demonstrate that the intended assignee (or subtenant) is of sufficient covenant strength to meet his obligations; and
- to give the landlord information about the terms of the transaction.

Information on pre-conditions

5.35 Pre-conditions on assigning and subletting are fairly common, especially in more modern leases. They are explained at 1.36 to 1.40. For example, the lease might require any subletting to be at market rent, or might require an assignee to provide a guarantor. If the tenant cannot comply with the condition, the landlord will not be obliged to deal with the application.[300]

5.36 If the condition is valid, the tenant should provide enough background information to show that the condition will be met. The tenant should state, for example, the proposed amount of the sublease rent, or provide details of the proposed guarantor.

Even for pre-conditions, the court has held that an application is not invalidated if the tenant fails to provide details of the sub-rent at the outset, but the tenant would have to provide the information if the landlord asked for it.[301] We consider that the same will apply to all types of pre-condition. It will therefore help speed up the process to provide it from the outset.

5.37 However, it may not be wise for the tenant to go further and state that he considers that the pre-condition is therefore satisfied. This is not necessary and, by specifically drawing the pre-condition to the landlord's attention, may alert him to a possible source of objection which might otherwise be overlooked.

Information on covenant strength

For further information on covenant strength as a ground for refusal, see 7.51 to 7.69.

5.38 The tenant should aim to show that the assignee or subtenant will be able to perform the covenants under the lease (or sublease). The tenant should therefore provide the following:

- Three years' accounts for the assignee or subtenant. These should ideally show pre-tax profits of at least triple the rent, although the courts have warned against the slavish application of this general rule,[302] so it is not a strict legal requirement (see 7.67 to 7.69).

- A business plan, or some other explanation of the assignee's or subtenant's business. This will support the application, especially if it tends to show a well-run or growing business or if the assignee or subtenant is a new company or does not meet the 'test' of profits exceeding triple the rent.[303]

- A comprehensive set of references. The following points should be noted:
  - The set should include a landlord's, accountant's and trading reference where appropriate.
  - The letters seeking the references should also be enclosed, unless the reference itself includes that information.[304]
  - Referees' standing and qualifications to give the reference should be made clear where appropriate.[305]
  - Valuations of property should reveal the extent of any lending secured against the property.[306]
  - References should make clear the facts on which they are based; for example, trading references should state the level of business done with the assignee[307] and banking references should reveal any substantial or unusual liabilities.
  - Information about a person's income should make clear whether the figures are before or after tax.[308]
  - Disclaimers in references detract from their weight.[309]

– If the assignee or subtenant is taking on a greater
liability than he has had before, an accountant's
reference should confirm that the assignee or
subtenant can manage the extra burden, and
explain the reasons for that opinion.[310]

Information on the transaction

5.39 The landlord is entitled to know the nature of the
transaction.[311] It is therefore advisable for the tenant to
enclose general details of the structure of the assignment
or subletting with the application. This is often done by
providing a copy of agreed heads of terms.

5.40 The landlord's entitlement to know the nature of the
transaction also covers any side letters or collateral
agreements.[312] Therefore, it is not acceptable for the
tenant superficially to comply with any conditions on
assigning or subletting but simultaneously enter a side
agreement to breach them.

Example

A tenant has a lease of an industrial unit with ten years remaining
and wants to dispose of the premises. The lease requires sublettings
to be at market rent, and to mirror the rent review provisions in
the lease. To comply with this, the sublease would have to contain a
rent review in five years' time. A subtenant is found. It agrees to pay
the market rent but is only prepared to take the sublease if there
are no rent reviews.

The tenant applies for consent. He encloses references and three
years' accounts for the subtenant, and states the amount of the
sublease rent. However, he tries to avoid the requirement for a rent
review by using a sublease that appears to comply, but with a side
letter. The side letter negates any increase in rent following the
review.

The landlord is entitled to take the terms of both the sublease and
the side letter into account. The condition in the lease has
therefore not been satisfied and the landlord can refuse to
consider the application.

5.41 However, the landlord does not necessarily have a right
to see the detailed terms: he is entitled to know about

anything in the transaction that will affect him, but no more. So, on an assignment the landlord is not entitled to know the terms of the assignment or any premium or reverse premium, unless they affect him. But the terms of a subletting *will* affect the landlord, so he is entitled to see the proposed sublease.[313]

Alterations

5.42 On an application for consent to alter, the tenant must make sufficiently clear what his proposals are that the landlord knows whether to give or refuse consent.[314]

5.43 The tenant should therefore enclose with the application clear and comprehensive plans and specifications for the works. If the tenant is a business tenant and is able to use the special scheme for improvements, see also 3.76 to 3.93. This may be relevant if the tenant wants to claim compensation later, as well as if the tenant is trying to overcome an absolute restriction.

5.44 In our opinion, the following might also speed up the application, or make it more difficult for the landlord to refuse:

- if the tenant has engaged reputable builders or architects to work on the project, include their details;
- nominate an individual with detailed knowledge of the plans who will be available to answer any questions the landlord (or his advisers) might have;
- if appropriate, include a statement from a qualified person that the works will not affect the structural integrity of the building;
- if the tenant has obtained planning permission for the works, enclose a copy of the permission; and
- if the tenant believes that the works might reduce the value of the landlord's interest, make an offer to pay reasonable compensation. The tenant might propose a sum but offer to pay whatever amount a court deems reasonable. For more information on compensation as a condition of landlord's consent, see 3.46 to 3.49 and 7.152 to 7.153.

Changes of use

5.45 The tenant should specify in detail the nature of the proposed new use. Aside from that, the following might make it more difficult for the landlord to refuse consent:

- if the tenant has obtained planning permission for the change of use, enclose a copy of the permission; and

- if the tenant believes that the change of use might reduce the value of the landlord's interest, make an offer to pay reasonable compensation. The tenant might propose a sum but offer to pay whatever amount a court deems reasonable. For more information on compensation as a condition of landlord's consent, see 4.55 to 4.57 and 7.180 to 7.181.

5.46 If the restriction is qualified, but not fully qualified (i.e. there is no requirement that consent should not be unreasonably withheld), it would also be useful to state (if appropriate) that there are no structural alterations involved. This is because the landlord is entitled to ask for a premium if the change of use involves structural alterations:[315] see 4.53 to 4.60.

Costs

5.47 The landlord will probably want to ask the tenant to pay his costs for dealing with the application. The tenant should therefore check whether he will be liable to meet them.

5.48 The first step is to check the lease to see if it provides an answer.

- If the lease is silent, the landlord is probably entitled to demand a solicitor's undertaking to pay the landlord's reasonable costs before dealing with an application. However, in an application for consent to sublet, the courts have held that it is unreasonable to ask for an undertaking for *all* costs, rather than just *reasonable* costs.[316] Presumably the same would apply to other types of application.

- If the lease contains a tenant's covenant to pay the landlord's costs, it may be unreasonable to ask for an undertaking at all, at least if the landlord already has satisfactory security for the costs[317] and perhaps if the tenant is a strong covenant. Alternatively, the lease may expressly require an undertaking as a pre-condition of the landlord's obligation to deal with an application.

5.49 If a tenant wishes to minimise a landlord's opportunities to delay, it would be sensible to confirm, in the application, that the landlord's reasonable costs will be met. If the application is made by the tenant's solicitors, this confirmation can be given in the form of a professional undertaking.

5.50 For further detail on costs, see chapter 8.

Urgency or importance to the tenant

5.51 If there is any special reason why the application is particularly urgent, it will do no harm for the tenant to spell that out. For example, in the case of an assignment, does the transaction form part of the sale of a business? Is it part of a larger series of transactions? In the case of alterations, do the works form part of a larger project which is already underway?

5.52 These kinds of factors may influence the amount of time a landlord can reasonably take to make a decision. They may also tip the balance in favour of granting consent, if there is a disproportion between the benefit to the landlord and the detriment to the tenant if consent is withheld.[318]

> For more information on how long the landlord can take to respond, see 6.26 to 6.29.

Serving the application

5.53 Once the tenant has prepared an application, care should be taken to serve it in the correct way, and at the correct address.

5.54 The approach will differ if the tenant is making an application for consent to assign or sublet, compared with any other type of application.

5.55 Note that, as explained at 5.23 to 5.25, the application may need to be served on an RTM company (if one is in place) rather than the landlord.

5.56 If the landlord's agent or solicitor is aware of the impending application, there is no harm in sending a copy to him, provided the landlord is served.

Applications for consent to assign or sublet

5.57 The tenant will want to ensure that the application is properly served. This will trigger the landlord's duties under the LTA 1988.

For information on when the LTA 1988 applies and what the landlord's duties are, see 1.22 to 1.31.

5.58 The Act refers to two alternatives: where the lease specifies a method of service, and a fall-back provision if the lease is silent.[319] The tenant should therefore check the lease first. If a method of service is set out, he should use that method. If not, use the alternative.

Method of service required by the lease

5.59 The tenant should follow carefully any requirements in the lease. For example, does the lease specify that a particular address must be used, or that a reference must be quoted on any correspondence? Alternatively, the lease might specify that a particular method of service is not permitted. For example, it might state that applications may not be served on agents but must be served on the landlord personally.

5.60 If a requirement is not followed, the application may not be properly served, and the landlord may have no obligation to deal with it.

Method of service permitted by the lease

5.61 Sometimes the lease will not contain any positive requirements for service, but will give the tenant options for service. These are not compulsory, but if one of the options is used, the application may be deemed to have been served even if it never arrives.

Example

A lease states: *'notices may be served on the landlord at its registered office by first class post. Unless the notice is returned undelivered, it shall be deemed served on the second working day after posting.'* The tenant may serve in the way described, but he could choose to serve in a different way.

5.62 It is common for leases to incorporate the provisions on serving notices set out in s. 196 of the LPA 1925. Where they apply, these provisions give the tenant extra choices on how to serve the application.

5.63 If s. 196 is incorporated into the lease, the application can be served by leaving it at, or sending it by registered post to, the landlord's last-known 'place of abode or business'.[320] If the notice is sent by registered post, it is only served if it is not returned undelivered.[321] For a tenant of residential property, the last-known address includes an address given by the landlord for service of notices, or failing that, an address given in a rent demand.[322]

If the lease is silent

5.64 If the lease does not have any provisions relating to service, the tenant should follow s. 23 of the LTA 1927.[323] This allows the tenant to serve by:

- handing it to the landlord personally (in the case of an individual);

- leaving it at, or sending it by registered post to, the landlord's 'last known place of abode in England and Wales'; or

- in the case of a local or public authority, or certain special types of company, sending it by registered post 'to the secretary or other proper officer at the principal office of such authority or company'.

5.65 The Act expressly states that serving the notice on the landlord's agent will suffice, provided the agent is authorised to accept notices,[324] but as explained at 5.71, serving the landlord direct is the safer option.

Actual receipt and effect of service by a permitted method

5.66 These methods of service are not exhaustive,[325] so any application actually received by the landlord, or which comes to his (or his authorised agent's) attention, will be valid. This is explained further at 5.70 to 5.71.

5.67 However, the advantage of using one of the permitted methods is that the difficulties of proving service are much reduced. The application will be *deemed* to be served even if the landlord does not actually receive it.

Examples

(1) The tenant in the example above sends the application to the landlord's usual place of business by DX. It actually arrives, so it is served.

(2) The tenant sends the application to the landlord's registered office by first class post. It gets lost in the post. In accordance with the lease, it is deemed served.

(3) The tenant sends the application to the landlord's usual place of business by DX. It gets lost in the DX. It is not served.

Applications for consent to alter or change use

5.68 Unlike the case of applications for consent to assign or sublet, there are no statutory provisions to help the tenant.

5.69 Therefore, the tenant should first check the lease for any provisions about notices. If there are any, follow the guidance above in 5.59 to 5.63.

5.70 If the lease is silent, the tenant must ensure that the application is received by, or comes to the attention of, the landlord.[326]

5.71 Service on an agent (such as a solicitor or managing agent) will suffice, but only if the agent is authorised to accept it.[327] Managing agents will often be taken to be authorised,[328] but the same is not true of solicitors.[329] Since the tenant will not normally know whether the agent is authorised, it is therefore sensible for the tenant to send or deliver the application to the landlord personally, or at an address where it is likely the landlord will receive it – for example, a known home, business property or (in the case of a company) the registered office.

After service of the application

5.72 If the landlord reasonably requests further information, the tenant should provide it. If the tenant fails to supply responses to reasonable enquiries and the landlord refuses consent, the refusal might be reasonable.[330] For more details on what further information the landlord can ask for, see 6.48 to 6.50.

Checklist: Making an application

Use this checklist to decide whether an application is needed and whether it will be properly served.

- Does the tenant need to make an application and does the landlord need to respond? See 5.1 to 5.6.
- Is the application in writing, sent by (or on behalf of) the tenant, and addressed direct to the correct person? See 5.7 to 5.27.
- Does the letter explain the application and include as much additional information as possible? See 5.28 to 5.46.
- Does the application include confirmation that the landlord's reasonable expenses will be paid? See 5.47 to 5.50.
- Will the application be served in the correct way? See 5.53 to 5.71.

Sample letters

The following sample letters show how the principles explained above might be put into practice.

There are letters dealing with each of the three types of restriction. If the tenant is making more than one application, there is no reason why they cannot be amalgamated into a single letter.

Letter seeking consent to assign the whole premises

For the purpose of this letter, it is assumed that:

- the landlord is a company;
- the lease incorporates the rules on serving notices in s. 196 of the LPA 1925 (see 5.62 to 5.63); and
- the tenant's surveyor is making the application.

[Tenant's surveyor's headed paper]

To:

[Landlord's full name
and business address]

[Date]

Dear Sirs

123 High Street, Anytown ('the Premises')

We act for Oldco Ltd. Oldco is your tenant of the Premises under a lease dated 7 July 2007 granted by High Street Holdings Ltd to High Street Shops Plc ('the Lease').

Oldco wishes to assign the whole of the Premises to Newco Ltd. On behalf of our client, we therefore seek your consent to the assignment pursuant to clause 3(8) of the Lease. To assist you in considering this application, we enclose the following:

(a) three years' audited accounts for Newco Ltd (which, you will see, demonstrate that Newco's pre-tax profits exceed triple the rent due under the Lease);

(b) a copy of Newco's business plan, which explains the nature of its business and how it is run;

(c) a set of references, together with the letters seeking the references, from each of the following:

    (i) an existing landlord of Newco;

    (ii) Newco's accountant; and

    (iii) Newco's main trading supplier; and

(d) a copy of the heads of terms for the proposed assignment. We confirm that there is no intention to enter into any collateral or side agreement.

We confirm that Oldco will meet your reasonable expenses for considering this application.

Newco is anxious to complete the assignment as soon as possible and in any event by this time next month. We therefore look forward to hearing from you as soon as possible.

Yours faithfully,

Letter seeking consent for alterations

For the purpose of this letter, it is assumed that:

- the landlord is an individual;
- the lease contains no requirements for how notices should be served; and
- the tenant is making the application itself.

[Tenant's headed paper]

To:
*[Landlord's full name
and address]*
[Date]

Dear Sir

123 High Street, Anytown ('the Premises')

We are your tenant of the Premises under a lease dated 7 July 2007 granted by High Street Holdings Ltd to High Street Shops Plc ('the Lease').

We wish to carry out alterations to the Premises, and seek your consent to do so pursuant to clause 3(7) of the Lease. In general terms, the works will involve making a new door and window in the front of the Premises, removing a wall in the ground floor, and installing a new staircase to the first floor. To assist you in considering this application, we enclose the following:

(a) a complete set of our architect's plans and specifications for the works;

(b) a letter from our architect confirming that in his opinion the works will not affect the structural integrity of the Premises; and

(c) a copy of a letter from the local authority granting planning permission for the works.

Our architect is Mr Smith of Smith Jones & Co. He would be happy to answer any questions you, or your advisers, might have about any aspect of the works. His details are set out in the enclosed letter.

We confirm that we will meet your reasonable expenses for considering this application.

We look forward to hearing from you as soon as possible.

Yours faithfully,

Letter seeking consent for a change of use

For the purpose of this letter, it is assumed that:

- the landlord is a company;
- the lease contains no requirements for how notices should be served; and
- the tenant is making the application itself.

[Tenant's headed paper]

To:

[Landlord's full name
and registered office
or other known business address]

[Date]

Dear Sirs

123 High Street, Anytown ('the Premises')

We are your tenant of the Premises under a lease dated 7 July 2007 granted by High Street Holdings Ltd to High Street Shops Plc ('the Lease').

We wish to change the use of the Premises from that of travel agency to a restaurant. We seek your consent to that change of use pursuant to clause 3(9) of the Lease. To assist you in considering this application, we enclose the following:

(a) a copy of our business plan giving full details of the nature of the proposed restaurant business; and

(b) a copy of a letter from the local authority granting planning permission for the change of use.

We confirm that we will meet your reasonable expenses for considering this application.

We look forward to hearing from you as soon as possible.

Yours faithfully,

6

Responding to an application

This chapter explains:

- the preliminary checks a landlord should make before responding;
- the duty to pass on the application;
- the landlord's duty to respond and how long he may take;
- the types of information the landlord may ask for; and
- the need to avoid inadvertently giving consent.

Before responding: Preliminary checks for landlords

6.1 Landlords often begin dealing with applications when there is no need. The proposed action may be completely prohibited, and if the landlord does not appreciate this he may fail to make the most of his position.

6.2 Conversely, the landlord may fail to deal with an application when he *does* need to. This can have serious consequences for him: he may lose his right to prevent a tenant's action, and he may even have to pay compensation as a result.

6.3 Therefore, before beginning to deal with an application, landlords should check the restrictions in the lease by following the checklist below. The process is similar to the one that tenants should follow before preparing an application (see 5.1 to 5.6).

Inttial checks

6.4   Before you begin:

- Check the lease and any ancillary documents such as a previous licence to find out whether the intended action is expressly restricted.

- Check the relevant section of this book (2.36 to 2.78 on alienation, 3.27 to 3.34 on alterations and 4.49 on change of use) to see if a restriction will be implied.

- If there is a restriction, does it cover the intended action? For example, there may be a restriction on subletting *part* of the demised premises, but is the tenant free to sublet the *whole*? Similarly, there may be a restriction on structural alterations, but do the intended works actually constitute 'structural alterations'?

- Does the restriction prohibit the intended action completely, or can the tenant go ahead with the landlord's consent? See 1.6 to 1.15.

- Does the lease expressly say that consent is not to be unreasonably withheld (i.e. is it fully qualified)? If not, check the table at 1.19 to see if this will be implied.

- Finally, check that the tenant has made a proper application (see chapter 5), and that any valid pre-conditions will be met (see 1.36 to 1.40).

Do you need to respond?

6.5   Based on those checks, the landlord can decide whether and how to respond:

- If there is no express or implied restriction that covers the intended action, the tenant may go ahead without consent and need not have applied. For example, if the lease does not restrict assignments, the tenant is free to assign the lease to anyone without adverse consequences. The application is therefore irrelevant and the landlord may choose whether or not he wishes to respond. There may be good reasons for responding if the tenant has not realised the strength of his position.

- If the tenant has not made a valid application or has not met a pre-condition, there is no need for the landlord to respond. The landlord should, however, be very confident of his analysis before choosing not to respond.

- If there is a restriction that completely prohibits the action, the tenant may ask the landlord to consent, but this will not be governed by the lease. The landlord may refuse consent for any reason and need not even respond at all.

  There is an important exception for alterations. If the tenant is using the scheme for improvements to business tenancies, the landlord must reply within three months, or the tenant may be entitled to go ahead anyway. The tenant's application may not make clear whether he is using this scheme, and the landlord would be wise to respond in all cases of doubt.

- If there is a qualified restriction that the law does not treat as fully qualified (i.e. qualified covenants against changing use and limited other covenants), the position is the same as a complete prohibition.

  However, in most cases other than change of use involving structural alterations, the landlord may not demand a premium.

See Appendix 2 for details of which covenants are not made fully qualified.

For details of when a premium is prohibited, see 2.79 to 2.91, 3.35 to 3.51 and 4.50 to 4.64.

- If there is a fully qualified restriction (i.e. the landlord's consent must not be unreasonably withheld), the landlord must act reasonably (and may have other duties as well).

Dealing with the application

6.6   If the landlord is obliged not to withhold consent unreasonably, and the tenant has made a valid application, the following sections explain how to respond.

6.7 This chapter only deals with how quickly the landlord must respond, and other practical issues involved with giving an effective response. The grounds on which consent can reasonably be withheld and conditions which can reasonably be imposed are dealt with in chapter 7.

6.8 A landlord should also consider whether there are any restrictions that prevent him from giving consent.

For example, a landlord will often need his own landlord's approval (explained at 1.53 to 1.54). See also 1.61 for information on the other types of covenants that might affect the tenant's plans.

If a superior landlord's consent is required, there are some important additional responsibilities. These are explained in the following paragraphs. See also 7.29 to 7.31 on using this as a ground for refusing consent.

If the landlord will breach some other covenant by granting consent, he should consider whether he can reasonably refuse consent on this basis. See 7.32 to 7.33. It will usually be wise to take legal advice in this situation.

Superior landlords

6.9 There are some additional responsibilities on landlords and superior landlords dealing with an application for consent to assign, sublet, charge or part with possession of the property. These duties are imposed by the LTA 1988.

See 1.27 to 1.31 for an explanation of when the LTA 1988 applies.

Passing the application 'up the chain'

6.10 The landlord may have an obligation to pass the application 'up the chain' to any superior landlord (in addition to his obligation to respond to the tenant). This is most likely to arise when the lease requires the tenant to obtain the landlord's *and superior landlord's* consent.

6.11 The LTA 1988 requires the landlord to pass the application to anyone else who may consent to the transaction. The landlord need not pass it on to a person who he believes has already received a copy.[331] The circumstances in which someone else's consent is required are explained at 1.53 to 1.54.

Example

A sublease contains a covenant by the subtenant '*not to assign, sublet or charge the premises without the consent in writing of the landlord and superior landlord, such consent not to be unreasonably withheld*'.

If the subtenant applies to his immediate landlord for consent, the landlord must then pass a copy of the application to the superior landlord.

If the subtenant's covenant does not refer to the superior landlord, the subtenant is under no direct obligation to obtain his consent.[332] However, the landlord may still need to obtain the superior landlord's consent before giving his own approval: see 6.16 to 6.17.

6.12 The landlord must take reasonable steps to ensure that the superior landlord receives a copy of the application within a reasonable time.[333]

6.13 The courts have not given any guidance on how long the 'reasonable time' might be in this context. However, we suggest that it would probably be unreasonable for a person to take more than a week to send the application on, unless there are any reasonable difficulties in identifying the intended recipient or finding the correct address. It might be acceptable to take a little longer in August or at Christmas.[335]

6.14 Although the duty to pass on applications will usually fall on landlords, it also applies to anyone else who receives the application and is a person who may consent to the transaction.[334] This will catch superior landlords whose consent is required, who must then pass the application 'down' to the immediate landlord, and RTM companies.

Superior landlords' duties

Duty to respond

6.15 If a superior landlord receives an application for consent to dispose of the property, he has the same duties as the tenant's immediate landlord.[336] These duties are explained further at 6.20 to 6.29.

Duty to approve consents given by tenant

6.16 The superior landlord might also be called on to approve a consent given by his immediate tenant to a subtenant. For example, the immediate tenant might covenant '*not without the approval of the landlord (such approval not to be unreasonably withheld) to consent to the subtenant assigning or further subletting the demised premises*'.

6.17 In a situation where:

- the immediate tenant covenants not to consent to the subtenant assigning or subletting without the approval of the superior landlord; and
- the superior landlord's approval is not to be unreasonably withheld; and
- the superior landlord is served with an application for consent

the superior landlord will have similar duties to the tenant's immediate landlord. Broadly speaking, this means that within a reasonable time he must give approval (unless he can reasonably withhold it) and must serve notice of his decision either way.[337] He must give notice both to the tenant and to the subtenant.[338] These duties are explained in the following paragraphs.

Landlords' duties

6.18 The landlord's position will be different depending on whether he is faced with:

- an assignment or subletting case; or
- an alterations or change of use case.

6.19 These are explained separately.

Assignment or subletting

6.20 The landlord must respond within a reasonable time Within that time, he must:

- give consent (unless it is reasonable to withhold it); and

- in any case, serve written notice of his decision on the tenant, including any conditions on the consent or the reasons for withholding it.[339]

6.21 This comes from the LTA 1988. In rare cases where there is a fully qualified covenant against assignment etc. to which the LTA 1988 does not apply, the principles for alterations or change of use will apply.

For details of when the LTA 1988 applies, see 1.27 to 1.31.

For secure tenancies, see 6.30 to 6.31.

6.22 There are two aspects to giving an effective response:

- giving notice to the tenant; and

- doing so within a reasonable time.

Notifying the tenant

6.23 The landlord must notify the tenant of his decision.

6.24 His duty, according to the LTA 1988, is to serve written notice of the decision within a reasonable time, specifying in addition any conditions attached to the consent and, if consent is withheld, the reasons for withholding it.

6.25 There are several important principles that flow from, and affect, this duty.

- First, the landlord must give reasons for a refusal, or specify any conditions attached to his consent.[340] A consent given subject to an unreasonable condition is treated as an unreasonable refusal.[341]

- Secondly, the landlord is confined to the reasons given in the written response.[342] He cannot later justify the decision by reference to reasons not stated, even if they actually influenced the decision and were given orally to the tenant within the reasonable time.[343] However, this rule does not prevent a landlord from explaining the reasons he put in his written response or from relying on evidence to supplement them.[344]

- Thirdly, the landlord cannot rely on matters which did not influence him at the time he refused consent.[345] The landlord cannot therefore justify the decision by facts which later come to light, no matter how relevant they might be.

- Fourthly, if the landlord relies on both good and bad reasons for a refusal, he will have acted reasonably. The presence of a bad reason will not usually vitiate a refusal based on a sound independent reason.[346]

- Fifthly, once a landlord has given a decision the reasonable time comes to an end (even if the landlord would otherwise have been entitled to more time). The landlord cannot then reconsider his decision; if it was unreasonable he will still be bound by it.[347]

- Finally, if the landlord does not respond within a reasonable time, he may not rely on any reasons at all, and the withholding is automatically unreasonable, even if there actually are good reasons to withhold consent.[348]

How long is a 'reasonable time'?

6.26 One of the questions of greatest concern to both landlords and tenants is 'how long is a reasonable time?' This is because failure to give a decision within a reasonable time is counted as an unreasonable refusal. This will make the landlord liable to compensate the tenant for any loss it suffers as a result (explained further at 10.67 to 10.79). The landlord cannot avoid this liability even if it can show subsequently that there were reasonable grounds for refusal.[349]

6.27 Unhelpfully, there is no certain answer. It will depend on the particular facts in every case. However, the courts have developed the following principles, which may be of some assistance:

- The reasonable time begins when the tenant makes a valid application for consent.[350] If there have been informal exchanges between the landlord and the tenant, the process will only begin when the tenant puts the application on a formal footing.[351]

- The reasonable time can only be judged at the end of the period, in the light of the facts at that time.[352] It can therefore be affected by events occurring after the tenant makes the application. For example, a failure by the tenant to deal promptly with reasonable requests for information may extend the reasonable time.[353]

- Any urgency (or lack of it) that the tenant communicates to the landlord can be taken into account.[354]

- If the landlord reasonably seeks further information, he may need to be ready to make a decision very quickly (perhaps within a week) once everything is to hand.[355]

- In an application that was not uncomplicated and which 'raised unusual legal and estate-management issues that merited serious consideration', a period of less than three weeks after receipt of all the information was not inherently unreasonable.[356] The court also took into account the fact that the events took place in the summer holiday season.

- Even in complicated cases, the reasonable time should usually be measured in weeks rather than months. A period of almost four months is unlikely ever to be acceptable, except perhaps in the most unusual and complex cases.[357]

- As stated above, the reasonable time comes to an end as soon as the landlord gives his decision.

- It will not matter if the tenant, for pragmatic reasons, has co-operated in requests made by the landlord for further information after the reasonable period has expired. This is the case unless it can be said that the tenant has waived the landlord's delay, or otherwise acted in a way that makes it unfair for him to rely on it.[358]

6.28 It is dangerous to try to draw any general conclusions about how long will be 'reasonable', but landlords should probably be concerned if a month, or perhaps less, has passed from receipt of the application. This is especially so if the application included a comprehensive set of information, and the tenant has dealt with any further requests promptly.

Example

A tenant has a lease of a warehouse unit. It wishes to sublet, but the lease contains various difficult pre-conditions, so it approaches the landlord and there are several informal discussions. Eventually, the tenant finds a possible subtenant and negotiates a complex deal. The landlord is informed, but little information is offered and he has to ask for further details, which take a long time to arrive. Three weeks after agreeing the subletting, the tenant accuses the landlord of exceeding the reasonable time.

It is unlikely that the reasonable time has expired. The arrangement is complex, the tenant never told the landlord that the application should be examined formally, and they took a long time to deal with the requests for information. It is therefore unlikely that the landlord would be regarded as having had sufficient time to consider the application.

Practical points

6.29 The principles above inform how the landlord should prepare a response. The following points may be of practical help:

- The landlord has the difficult task of gauging how long the 'reasonable time' will be in each case. Any unusual features or extra complexity in the transaction are likely to lengthen that time.

- Bear in mind that the landlord may ask for reasonable further information (see 6.48 to 6.50). If the tenant does not respond promptly, the reasonable time will be extended.

- The landlord should give a written decision in every case, within the reasonable time.

- The response should set out all the reasons on which the landlord relies. Good and bad grounds are discussed in chapter 7.

- The landlord should try not to 'jump the gun' and give a decision too soon, since he cannot then revisit it in the light of any new information (unless, perhaps, the lease contains an express clause allowing him to do so).

Example

A tenant wishes to sublet an office building. The landlord refuses consent, stating that he is concerned about the subtenant's financial strength, and because in any case he wants to vary the terms of the lease. He is also concerned that the subtenant might breach the user clause, but forgot to mention that in his letter.

If the tenant accuses the landlord of unreasonably withholding consent, the only reasons to be taken into account are the concern about financial strength and the intended variation of the lease. If the concern about financial strength is valid, that will be a reasonable ground for refusal (and will not be vitiated by the reliance on an unreasonable one: the wish to vary the lease). The concern about the use of the premises is irrelevant because it was not stated in the refusal letter.

Secure tenancies

A secure tenancy is a type of residential public sector tenancy. For further details, see Appendix 1.

6.30 The LTA 1988 does not apply to secure tenancies.[359] However, if a secure tenant applies to exchange his tenancy (see 2.41), the landlord must respond in writing within 42 days. As set out in 7.185 to 7.186, he may only rely on certain grounds for refusal and impose certain conditions. The notice must specify the ground and give particulars of it. If the landlord misses this deadline, consent will be treated as having been given.[360]

6.31 In addition, if a secure tenant applies to sublet or part with possession of part of his dwelling-house, the landlord must respond within a reasonable time. If consent is refused he must give the tenant a written statement of reasons why it was refused.[361]

Alterations or change of use

6.32 In an alteration or change of use case, the tenant does not have the added weaponry of the LTA 1988. Therefore, with the exception of alterations to some residential properties or those made to comply with the *Disability Discrimination Act* 1995 (see 6.36 to 6.41), the landlord is not under any special duty to consider the application within a reasonable time, and most of the guidance at 6.23 to 6.29 does not apply.

6.33 The obligations on the landlord, and the consequences if he fails to comply with them, are much more limited. The only avenue for the tenant is to say that the landlord has unreasonably withheld consent. There is little guidance from the courts on how long a landlord can take for this purpose.

6.34 The following principles can be drawn from the decisions of the courts, but ultimately the question will simply be whether, in the opinion of the court, the landlord's conduct amounts to unreasonable withholding of consent.

- The tenant is entitled to have an application dealt with 'expeditiously'. Failure to do so is an unreasonable withholding of consent.[362] The landlord should not be allowed to take so long that the object of the application is defeated.[363]

Example

A landlord was found to have withheld consent unreasonably where there had been lengthy correspondence over about five and a half months, he was still requesting further information and had not given a decision.[364]

- The landlord is not obliged to give reasons for the refusal.[365] However, a failure to give reasons may mean that a court is more willing to conclude that the refusal was unreasonable.[366]

  This is a problem for tenants. They may have no way of knowing whether the landlord has been reasonable or not, making it hard to decide what action to take in response.

- Unlike assignment and subletting cases, the landlord is not confined to the reasons which were given to the tenant at the time of refusal. The landlord can therefore justify the decision with other unstated reasons,[367] provided the reasons actually influenced the decision.[368]

- An unreasonable withholding of consent will not usually make the landlord liable to the tenant for damages. The tenant's only options are to go ahead with the alterations or change of use,[369] or to apply to court for a declaration that he may do so.

  This is because a tenant's covenant not to assign without consent, such consent not to be unreasonably withheld, does not amount to a covenant by the landlord.[370] However, if the wording is unusual, there might be an express covenant by the landlord not to withhold consent unreasonably; if he breaches that obligation, he will be liable to pay damages.[371] This is explained further in chapter 10.

Some of these principles come from cases on alienation covenants, but we consider that they should apply equally to alterations and changes of use.

Example

A tenant decides to carry out some major alterations to its warehouse premises. It wishes to do structural works to convert part of the warehouse into office space, and applies to the landlord for consent. The landlord asks for information about the effect of the works on the structure of the warehouse. This is provided. Then, after further correspondence, the landlord withholds consent without giving any explanation.

The request for information about the effect of the works is likely to be reasonable (see 6.47 to 6.49). The failure to give grounds for the refusal is not, of itself, unreasonable. If the landlord can show that he was influenced by reasonable factors, he will not have unreasonably withheld consent. However, the tenant may be able to use the special scheme for improvements: see 3.76 to 3.93.

Improvements to properties let on business tenancies

6.35 Where the tenant is using the special scheme for improvements described in 3.76 to 3.93, the landlord must respond within three months. If he fails to do so, the tenant will be entitled to go ahead. The tenant's application may not make clear whether he is using this scheme, and the landlord would be wise to respond in all cases of doubt.

Alterations under the Disability Discrimination Act 1995

6.36 The law imposes a special regime on applications for works required under the disability laws.

6.37 The circumstances in which this regime applies are explained at 3.56 to 3.60. It covers works required to comply with:

- the duty on employers to make reasonable adjustments to premises;[372] and

- the duty on service providers not to discriminate against disabled customers.[373]

6.38 In either case, the landlord of the employer or service provider is taken to have withheld consent if he does not respond within a certain period, either consenting to or refusing the application, or consenting subject to obtaining a superior landlord's (or any third party's) consent. The landlord must, however, seek that consent within the deadline.

6.39 The periods for landlords to respond are:

- for the duty to make reasonable adjustments: 21 days (or such longer period as is reasonable) from receipt of the tenant's application;[374] or

- for the duty not to discriminate against disabled customers: 42 days from receipt of the tenant's application.[375]

6.40 A landlord who fails to comply with this duty is treated as having withheld consent unreasonably.[376]

6.41 There are detailed regulations setting out other requirements and time limits that the landlord must comply with.

Secure and Rent Act tenancies

A secure tenancy is a type of public sector residential tenancy. Protected and statutory tenancies are types of private sector residential tenancy usually entered into before 15 January 1989. For further details, see Appendix 1.

6.42   Some residential tenancies are subject to a special regime affecting applications for consent to carry out alterations. This regime applies to secure, protected and statutory tenancies. All of these types of tenancy are deemed to contain an implied term that tenants may carry out certain alterations with the landlord's consent, which is not to be withheld unreasonably. This is explained at 3.31 to 3.34.

6.43   Where this regime applies, the landlord will be taken to have withheld consent where he fails to give or refuse consent within a 'reasonable time'. The landlord must give the tenant a written statement of reasons if consent is refused.[377]

RTM companies

6.44   Long leaseholders of flats have the right to take over management of their building through an RTM company, which then takes over most of the landlord's functions relating to consents. There are special rules about how RTM companies must deal with applications for consent. This is explained at 5.23 to 5.25.[378]

6.45   The RTM company is prohibited from granting consent to an application by the tenant without first giving notice in writing to the landlord. The required notice periods are:

- for an application to assign or underlet, do structural alterations or improvements, or change use: 30 days; and
- in other cases (for example, alterations which are not 'structural'): 14 days.[379]

6.46   If the landlord does not object within the notice period, the RTM company will be entitled to grant consent.

6.47   If the landlord does object, the RTM company may not grant consent unless the landlord later agrees in writing, or the Leasehold Valuation Tribunal makes an order.[380] The landlord's objection must be in writing, and be sent to the RTM company and the tenant (and subtenant if applicable).[381] The objection must be on valid grounds, applying the normal principles (as though the application had been made to the landlord).[382]

Summary: How quickly must the landlord respond?

This is a summary only, and reference should be made to the full explanations given above.

	Landlord's time limit	Obligation to give reasons?	May landlord rely on reasons not given?
Assignment and subletting	Duty to respond within a 'reasonable time': measured in weeks not months	Yes, the landlord must serve written notice of the decision, with reasons	No
Exchange of secure tenancy	42 days	Yes, the landlord must serve written notice of the grounds, with particulars	No
Alterations and change of use (except special cases)	Not defined: the landlord must not unreasonably withhold consent	No, but more likely to be unreasonable if no reasons are given	Yes, provided they did influence the decision
Alterations under disability laws	Either (a) 21 days (or longer if reasonable) or (b) 42 days, depending on the situation	No express obligation to do so	Answer is unclear
Alterations to secure, protected and statutory tenancies	Duty to respond within a 'reasonable time'	Yes, the landlord must give a written statement of reasons for refusing	Answer is unclear
Notice from RTM company	Within 14 or 30 days, depending on the situation	The landlord must give written notice of the objection, and any conditions attached	Answer is unclear

Asking for further information

6.48 The landlord may want to ask the tenant for further information about some aspects of the application. However, care should be taken to avoid going too far, and asking for information which the courts do not regard as reasonably necessary. Insisting that the tenant provide an unreasonable degree of information is likely to mean that the landlord has unreasonably withheld consent.

6.49 The following principles should be borne in mind:

- *Onus on landlord:* At least in the case of an assignment or subletting, the landlord must positively seek any further information he requires, and must ask in clear terms. If he does not ask, he cannot complain that the tenant has failed to provide it.[383]

- *Costs:* The landlord should check the lease to see if it deals with payment of costs and whether the landlord can demand an undertaking. If the lease contains a tenant's covenant to pay, but not a covenant to provide a solicitor's undertaking, it may not be reasonable to demand an undertaking before dealing with the application, at least if the landlord already has security which will cover the costs, and perhaps if the tenant is a strong covenant.[384] Even if the lease is silent, it will be unreasonable to ask that the undertaking extends to all costs, rather than just reasonable costs.[385] See chapter 8 for further information.

- *Pre-conditions:* It is reasonable to ask the tenant to demonstrate that any pre-conditions in the lease (for example, that any sublettings must be at market rent) will be met.[386] See 1.36 to 1.40 for more information about pre-conditions.

- *Nature of the transaction:* The landlord is entitled to know the nature of the transaction, including any side letters or collateral agreements.[387]

- *Subletting:* The landlord is entitled to see the proposed sublease.[388]

- *Assignment:* The landlord is not entitled to know the detailed terms of the assignment or the amount of any premium or reverse premium,[389] unless this is relevant to a pre-condition. It is therefore probably unreasonable to ask.

- *Covenant strength:* It is reasonable for the landlord to want to satisfy himself of the proposed assignee's (or subtenant's) ability to pay the rent and perform the other covenants in the lease.[390] Therefore, enquiries genuinely directed at that question will be reasonable. For example, it is reasonable to be concerned about the details of inquiries made to referees and the referees' qualifications to give the references.[391]

  However, landlords should be wary about taking these enquiries too far: the courts have held that landlords should also take into account the surrounding circumstances.[392] For example, does the business appear to be well-run?

  There is less scope for a landlord to be concerned about covenant strength on a subletting than on an assignment (but see the discussion of covenant strength as a ground for refusal at 7.51 to 7.69).[393]

  It is not reasonable for the landlord to enquire about a proposed assignee's ability to pay the rent under any possible statutory renewal of the lease.[394]

  The landlord is not entitled to interview the proposed assignee where satisfactory references have been provided.[395]

- *Proposed use:* A landlord can reasonably object to an assignment or subletting on the basis of the proposed use of the premises.[396] It should therefore be reasonable for a landlord to enquire about the proposed use.

- *Alterations:* The landlord is entitled to know what is proposed to be done to the property, and to know the tenant's solution to any structural problems.[397]

Example

A tenant wants to assign its lease and applies for consent. It will be for the landlord to seek any further information he requires. He may ask for reasonable information, such as accounts and references, to check that the proposed assignee will be able to meet the obligations in the lease. He may also ask about how the assignee would use the premises. It would be unreasonable, however, to ask to know the terms of the assignment or the amount of the premium (or reverse premium) changing hands.

6.50 The landlord should be wary of indicating to the tenant that consent will be given if the further information he seeks is provided. In some cases (depending largely on the precise meaning of the correspondence) this has led the courts to deem that consent was given after the landlord confirmed receipt of the information.[398]

The risk of inadvertently giving consent

6.51 Landlords should take great care when responding to an application, to avoid inadvertently giving consent. There are two main traps for unwary landlords: giving consent in correspondence and relying on the phrase 'subject to licence'. These are explained at 9.17 to 9.23.

Checklist: Responding to an application

Use this checklist to find out whether the landlord needs to respond to the application, and if so what the other duties and pitfalls are.

- Does the landlord need to respond? See 6.1 to 6.5.
- If so, consider whether the landlord needs to pass the application to any superior landlord. See 6.8 to 6.13.
- Next, consider how much time the landlord may take to give a decision. See 6.26 to 6.47.
- Consider what further information is needed. In particular, has the tenant shown that it can meet any valid pre-conditions? See 6.48 to 6.50.
- Remember the need to avoid inadvertently giving consent. See 6.51.
- Consider whether there are any restrictions which will affect the landlord's ability to grant consent. For example, is a superior landlord's consent needed? If so, see 6.8.
- Then move on to chapter 7 to look at whether the proposed grounds for refusing consent are reasonable.
- Finally, consider when and how to notify the tenant. See 6.23 to 6.47. If there are any pre-conditions which will need to be satisfied later (for example, that a sublease must be on

the same terms as the head-lease), consider whether to state that consent is given only on the basis that they will be satisfied: see 1.38.

7

Refusing consent or imposing conditions

This chapter:

- explains when the landlord must act reasonably;
- explains the general approach to reasonableness;
- gives examples of reasonable and unreasonable grounds and conditions;
- explains the special rules that apply to secure, protected and statutory tenancies; and
- explains the special rules that apply to alterations under disability laws.

If you are interested in the reasonableness of any particular reason or condition, see the index to locate it within the rest of this chapter.

When the landlord must act reasonably

> For the meaning of absolute, qualified and fully qualified, see 1.6 to 1.15.

Absolute covenants

7.1   If the covenant is absolute, the landlord can refuse a tenant's application for consent to do something restricted by the covenant. He may rely on any reasons or no reasons at all, and can impose any conditions he chooses. He can often even demand a premium. There is no need to act reasonably.

Example

A residential tenant covenanted 'not to carry out any alterations, save that non-structural internal alterations may be carried out with landlord's consent, such consent not to be unreasonably withheld'. He applies for consent to carry out what he says are non-structural internal alterations, but on close inspection of the application it is clear that the alterations are in fact structural. The landlord can safely refuse consent.

The landlord should, however, be aware of the special scheme that applies to improvements by business tenants. When this applies, if the landlord does not respond within three months the tenant will be able to go ahead anyway: see 3.76 to 3.93.

Pre-conditions

7.2 Many covenants contain pre-conditions. If the conditions are not satisfied, the covenant will be absolute; if they are satisfied, the covenant will be qualified or fully qualified.

For more information on pre-conditions, see 1.15 and 1.36 to 1.40.

7.3 The landlord may always refuse consent on the basis that a pre-condition is not satisfied, whether or not it is reasonable.

Example

A tenant covenants not to sublet without consent (not to be unreasonably withheld), provided that any sublease must be at full market rent.

The requirement that subleases must be at full market rent is a pre-condition. If this is not satisfied (i.e. if the proposed sublease is at less than market rent), the landlord may refuse consent without being required to act reasonably. If it is satisfied, the landlord may not withhold consent unreasonably (and his other duties under the LTA 1988 will also apply).

Subtenancies: sub-rent to equal or exceed head-rent

7.4 A pre-condition that causes particular difficulties in practice is '*no subtenancy shall be granted at a rent less than the rent passing under the head-lease*'. This can create a problem for tenants who wish to sublet premises which are over-rented. There are several potential ways of solving the problem.

- The tenant and the subtenant sign a lease in which the sub-rent is equal to the head-rent and a collateral deed in which:
  - the tenant agrees to indemnify the subtenant against any rent liability that exceeds the agreed market rent; or
  - the tenant agrees to pay the subtenant a sum equal to the difference between the sub-rent and the market rent on each rent day; or
  - the parties agree to reduce the rent that will actually be payable.

  Depending on the precise terms of the head-lease and the agreement between the tenant and the subtenant, this will usually be ineffective. The lease and the collateral deed will be read together, and when read together will be usually be taken as the creation of a subtenancy at a lower rent.[399] Therefore, the pre-condition is not satisfied and the landlord can refuse consent.

- The tenant pays the subtenant a reverse premium equal to the difference between the sub-rent (which will be equal to the head-rent) and the market rent, with any necessary adjustment for early receipt.

  Again depending on the precise terms, this will usually be effective; the pre-condition will be satisfied and the landlord may not withhold consent unreasonably.[400]

7.5 A reverse premium may be unattractive to the tenant, particularly if there is a risk of the subtenant becoming insolvent. The tenant may therefore wish to adopt a more sophisticated avoidance mechanism.

Qualified covenants

7.6 If the covenant is qualified, it will often be made fully qualified. If it is not (i.e. qualified covenants against changing use and limited other covenants), then the position is similar to that for absolute covenants; the landlord need not act reasonably.

7.7 However, in most cases other than change of use involving structural alterations, the landlord may not demand a premium. If he does demand an unlawful premium, the tenant can use the remedies described in chapter 10.

See Appendix 2 for full details of when qualified covenants are transformed. For when a premium may be demanded, see 2.79 to 2.91, 3.35 to 3.51 and 4.50 to 4.64.

Fully qualified covenants

7.8 If the covenant is fully qualified (i.e. the landlord's consent is not to be unreasonably withheld), then the landlord must act reasonably, and may only impose reasonable conditions. The landlord may not demand a premium except in unusual circumstances. If he fails to act reasonably, or imposes unreasonable conditions, or demands an unlawful premium, the tenant can use the remedies described in chapter 10.

7.9 The rest of this chapter explains when the landlord is likely to be acting reasonably, and what conditions are likely to be reasonable.

Setting out in the lease when consent may be withheld

7.10 Usually, the question of reasonableness is a question of fact, to be determined by a court if a dispute arises. The parties cannot usually agree in advance what will or will not be reasonable.[401] Any attempt to do so usually has no effect.

7.11 However, the parties *can* agree what will be reasonable in the case of:

- assignments of non-residential 'new' tenancies;[402]
- change of use; and

159

- limited other covenants where the law does not transform a qualified covenant into a fully qualified one.

> For more details about agreements on reasonableness and when these exceptions apply, see 1.33 to 1.52.

7.12 In these cases, effect will usually be given to the parties' agreement. If the landlord withholds consent on a ground which the lease says is to be deemed reasonable, or imposes a condition which the lease says is to be deemed reasonable, he will usually be on safe ground, even if the court thinks that the ground or condition is unreasonable. However, in the case of 'new' tenancies, there may be an exception if the agreement is void under s. 25 of the LT(C)A 1995: see 2.90.

7.13 However, the landlord must take care that his reason falls squarely within what is authorised by the lease, because the courts will strain to disallow an unreasonable ground.

Example

A lease contained a covenant against changing use without landlord's consent, which was not to be unreasonably withheld. It stated that the landlord's consent should not be treated as being unreasonably withheld if the landlord considered the proposed use to be in conflict with good estate management.

The landlord withheld consent to an application to sublet part of the premises and to change the use of that part. It stated that the subtenant's business would be in conflict with good estate management because it would lead to an over-intensification of the use of the site, causing traffic congestion.

The court held that the landlord's reason was not based on the *use* being in conflict with good estate management. Instead it was an objection to any subletting at all, regardless of the use. The agreement on reasonableness could only apply if the objection was based on the *nature* of the proposed use. The landlord's decision was not based on this, so it could only be upheld if it was actually reasonable – which it was not.[403]

The general approach to reasonableness

Refusing consent

7.14   The general principles on when it is reasonable to refuse consent to an assignment were set out in *International Drilling Fluids v Louisville Investments (Uxbridge).*[404] They were supplemented in *Straudley Investments v Mount Eden Land,*[405] which also made clear that the same principles apply to subletting as to assignment. However, as the legal consequences are different, the application of the principles is likely to differ in cases of subletting.[406] These principles should also inform the general approach to applications to make alterations or change use.[407]

The principles are:

- The purpose of the covenant restricting assignment is to protect the landlord from having his premises used or occupied in an undesirable way or by an undesirable tenant or assignee.

- The landlord is not entitled to refuse his consent on grounds that have nothing to do with the landlord and tenant relationship in regard to the subject matter of the lease.

- The landlord does not have to prove that the conclusions which led him to refuse consent were justified if they were conclusions which in the circumstances might be reached by a reasonable man.

- It may be reasonable to refuse consent because of the intended use of the premises even if that use is not expressly prohibited by the lease.

- The landlord need usually consider only his own interests, but in certain circumstances a refusal of consent may cause disproportionate harm to the tenant, and in such circumstances a refusal may be unreasonable.

- Reasonableness is to be decided as a question of fact in all the circumstances.

- It will normally be reasonable to refuse consent if this is necessary to prevent the landlord's contractual rights under the lease from being prejudiced by the proposed assignment/subletting.

- It will normally be unreasonable to impose a condition that would increase or enhance the rights that the landlord enjoys under the lease.

Imposing conditions

7.15 In many cases, it may be possible for the landlord's objections to be overcome by a suitable condition. For example, a concern about the tenant's covenant strength could be overcome by a guarantee or a rent deposit. However, the landlord is not obliged to propose conditions if he has sufficient grounds to object outright; instead he may simply refuse consent. The tenant may then make one or more further applications, offering conditions which may overcome the landlord's objection.[408] In other cases, it may be disproportionate and therefore unreasonable to refuse consent outright. If so, the landlord must instead grant consent subject to a condition.[409]

7.16 Care should be taken when imposing conditions:

- If a condition goes further than is necessary to protect the landlord, it is likely to be unreasonable.[410]
- If multiple conditions are imposed, and one of them is unreasonable, it may be that the tenant is entitled to go ahead without even complying with the reasonable conditions and (in an assignment or subletting case) that the landlord is liable to compensate the tenant for any loss the tenant suffers as a result.
- If the condition is not sufficiently precise, it may give rise to problems later.

Timing

7.17 The landlord might unreasonably withhold consent by failing to make a decision in good time (even if that decision would have been reasonable). This is explained further at 6.26 to 6.47.

Decision-making process

7.18 Sometimes the landlord's decision-making process may be flawed in some way. This is only relevant if the flaw leads to the reason being a bad one or if it affects the genuineness of the decision. The focus is always on what the landlord's reason actually is, and whether that reason is unreasonable.

Example

A council refused an application for consent to change use on the basis that the new use would attract an undesirable element to the area, leading to vandalism and disorder. In reaching this conclusion, the councillors had relied on some factually incorrect information. The reason was nonetheless objectively a good reason, even once the facts were corrected. The refusal of consent was therefore reasonable.[411]

7.19 Sometimes the landlord will make his decision for two or more reasons. If one of them is a good reason, his decision will usually be upheld.[412] However, if a bad reason vitiates the good reason, or if the real reason for the refusal is a bad one and the good reason is merely a makeweight, the good reason will not assist the landlord.

Use of illustrations

7.20 In the following sections of this chapter we give examples of reasons and conditions, drawn largely from case law, to illustrate which reasons and conditions are likely to be reasonable and which unreasonable. However, these are only illustrations. Each case turns on its own facts, and a ground which was reasonable in one case may be unreasonable in another.

All types of application: Reasons for refusal

Collateral advantage

7.21 It will normally be unreasonable for a landlord to refuse consent in order to obtain a collateral or 'uncovenanted' advantage, that is, one that enhances his position under the lease.[413] It is irrelevant that the landlord is motivated by principles of good estate management.[414]

Examples

(1) Withholding consent to assign in order to coerce the tenant into surrendering the lease is unreasonable.[415]

(2) Withholding consent to a change of use because the landlord wants to secure an alternative new use is also unreasonable. See the example at 7.178.

See also 7.38 on conditions which secure a collateral advantage.

7.22 In alienation cases, refusing consent in order to gain a collateral advantage may have serious consequences. If the landlord deliberately breaches his obligations in a way which is designed to achieve a collateral benefit, he may be liable to pay exemplary damages in addition to any financial compensation: see 10.73.

Preserving the status quo

7.23 Landlords are often permitted to preserve the status quo where a change would harm their interests; they are not permitted to force a change to the status quo where a change would benefit them.

Landlords should therefore check the lease to see if what they object to is contemplated by the lease as it currently applies. If it is *not* contemplated and the landlord would suffer harm, it may well be reasonable to refuse consent. If it *is* contemplated, an attempt to prevent it is likely to be viewed as an attempt to obtain a collateral advantage and a refusal is likely to be unreasonable.

Examples

(1) A 25-year lease contained a break clause which could only be exercised by the original tenant. The original tenant assigned the lease with landlord's consent, but the assignee became insolvent. The lease was an 'old' tenancy, so the original tenant had to pay the rent when the assignee defaulted. The assignee applied for consent to assign the lease back to the original tenant so that it could exercise the break clause. This would immediately reduce the value of the landlord's reversion by about £6m.

It was reasonable for the landlord to refuse consent. The break right could not be exercised by an assignee, so as matters stood, it could not be exercised at all. By refusing consent to assign, the landlord was not seeking a collateral advantage that it was not entitled to under the lease; it was simply preserving the status quo. Instead it was the tenant that was seeking a collateral advantage, by trying to achieve indirectly what it could not achieve directly.[416]

(2) A tenant applied for consent to assign a residential flat to an assignee who intended to sublet rather than live there. The lease permitted him to do so with landlord's consent which could not be unreasonably withheld. The landlords refused consent on the basis that a series of sublettings could give rise to problems. For example, occupation of the flat by young people on short tenancies, whose habits and attitudes were different from the mainly middle-aged or elderly professional people in the other flats, might cause friction.

The refusal was unreasonable. The lease clearly envisaged sublettings, so the mere possibility of a subtenancy being granted could not be a ground for objection. The landlords would be able to object to particular sublettings in the future if they had good grounds for doing so. In effect, the landlords were seeking to obtain a collateral advantage by seeking to ensure that the flat was only assigned to an owner-occupier, which was not required by the lease.[417]

Joint application – another part will fail

7.24   If a tenant makes several conjoined applications that are 'inevitably connected', the landlord is entitled to consider them together. If one of those applications will fail, the landlord may reasonably refuse consent to all of them.

Example

A tenant operated a jewellery shop in Regent Street. It applied for consent to assign the lease to a bureau de change/travel agent. The scheme would involve not only the assignment itself, but also a change of use and alterations to subdivide one unit into two. Consent was needed for each of them, not to be unreasonably withheld. Consent could reasonably be withheld to the applications to change use and make alterations. The application to assign was intimately connected with those other applications. The landlord could therefore refuse consent to assign as well.[418]

Preserving the right to forfeit

7.25 Sometimes the tenant may already be in breach of covenant or in arrears of rent, and the landlord may be deciding whether to forfeit the lease or not. If the breach is a 'once and for all' breach as opposed to a 'continuing' breach (see 10.21 to 10.24), he may be worried that if he responds to an application, he will lose his right to forfeit.

See 10.16 to 10.32 on forfeiture, and when the right to forfeit will be lost.

7.26 We consider that this is not a good reason for withholding consent to assign or sublet. Instead, the landlord should grant consent stating expressly that it is without prejudice to any accrued right to forfeit which he might have, and that the consent is conditional on this stipulation being accepted by both the tenant and assignee or subtenant.[419] This will preserve the landlord's right to forfeit, even for once and for all breaches.

7.27 In the case of continuing breaches, there is no risk of waiving the right to forfeit. There is therefore no basis for withholding consent,[420] and no need for a condition.

7.28 We consider that the same will apply to applications to make alterations or change use, although the position is not clear.

Consent of superior landlord not obtained

For an explanation of when the superior landlord's consent is required, see 1.53 to 1.54.

7.29 This is unlikely to be a reasonable ground for refusal.

7.30 If a superior landlord's consent is required, and is unreasonably refused, a landlord may not rely on that decision to justify his own refusal of consent. This is so even if the landlord's own lease contains a covenant not to give a licence without the superior landlord's approval (and as such the landlord fears breaching his own lease). [121]

We consider that the position would be the same if the superior landlord had simply failed to give a decision, or if his consent had never been applied for. Likewise, if the superior landlord's consent is not in fact required, a refusal on this basis will be unreasonable.

7.31 However, the landlord may be able to protect his position with a suitable condition: see 7.41 to 7.44.

Granting consent will put the landlord in breach of another covenant

7.32 Sometimes granting consent will put the landlord in breach of another obligation. For example, a landlord might need someone else's consent, such as his banker. This has been held to be a bad reason:

> 'If it could be a good reason, then any tenant, who had taken a lease from a landlord who, at the time that the lease was granted, had no financial restraints, could, if that landlord sold his interest to a party which had financial restraints, or found itself, for some reason, under such restraints, might suddenly, without notice, find the lease becoming more onerous'. [422]

7.33 However, we consider that this will not always be the case. If, for example, the landlord had covenanted with a neighbour before the lease was granted not to permit what the tenant has applied to do, we consider that it would be reasonable to refuse consent (at least if the tenant knew about it before taking the lease).

Tenant has not supplied information reasonably requested

7.34 If the landlord requests information which is reasonably necessary for him to make a decision, and the tenant fails to provide it, a refusal of consent will usually be reasonable.

See 6.48 to 6.50 on what information the landlord can ask for.

Example

A tenant applied for consent to assign to a Delaware registered entity. No information was supplied as to the assignee's financial standing. If the landlord had asked for that information and it had not been supplied, a refusal of consent on the ground of covenant strength would have been reasonable.[423]

Tenant has not supplied information reasonably required but not requested

7.35 If, on the other hand, the landlord fails to ask for the information within a reasonable time, then the fact that the information is reasonably necessary will not assist him, at least in the case of alienation. The landlord has only himself to blame for the lack of the information, and a refusal of consent will be unreasonable.

We consider that the same principle is likely to apply to use and alterations covenants. This is because any reason on which the landlord wishes to rely must have actually influenced the decision. If the landlord fails to ask for information, it is doubtful he will prove that the issue did have such an influence.

Example

In the example above, the landlord did not in fact ask for information about the assignee's financial standing. The refusal of consent was therefore unreasonable.[424]

Tenant has not given an undertaking for costs

7.36 If the landlord has reasonably requested an undertaking for costs, and the tenant has failed to provide one, it will usually be reasonable for the landlord to withhold consent.

See 8.6 to 8.9 for further details on when a landlord can reasonably request an undertaking for costs.

Professional advice not to give consent

7.37　The mere fact that the landlord has relied on professional advice in reaching a decision does not render the decision reasonable. The question is whether the advice was reasonable.[425]

> #### Example
>
> A tenant had a lease of a nightclub, and applied to change the use to a gym and fitness club. The landlord was advised that the change of use would significantly affect the value of the reversion, and refused consent. In fact, the advice to the landlord had been hopelessly flawed and was unreasonable. This rendered the decision to refuse consent unreasonable.[426]

All types of application: Imposing conditions

Collateral advantage

7.38　Imposing a condition that seeks to secure a collateral advantage for the landlord is likely to be unreasonable.

> #### Examples
>
> (1) A refusal to consent to an assignment except on condition that the lease is varied is unreasonable,[427] unless it is a variation aimed specifically at overcoming a *reasonable* objection and goes no further than is necessary.[428]
>
> (2) A condition that the landlord should have control over a proposed subtenant's rent deposit account was unreasonable. The condition sought to impose a safeguard for the landlord additional to those already available to him.[429]

Payment of money

7.39　A demand for a premium will be unreasonable, unless it is a rare case in which the lease validly permits the landlord to demand a premium even though consent cannot be unreasonably withheld.

7.40 The landlord can reasonably demand:

- reasonable compensation for damage to the value of the property or any of his neighbouring properties in the case of applications to make alterations or change use; and

- reasonable costs of dealing with an application. The landlord may even make this a condition of dealing with the application at all, not merely of granting consent.

However, it may not be reasonable to impose a condition that the costs of a previous abortive application are paid, at least if there is a genuine dispute about those costs.[430]

See also 7.152 to 7.153 on alterations and 7.180 to 7.181 on change of use.

For details of when landlords can demand a premium, and how requests for compensation should be phrased, see 2.79 to 2.91, 3.35 to 3.51 and 4.50 to 4.64.

See chapter 8 for guidance on what constitutes a premium, and more information about costs.

Consent of superior landlord

7.41 As explained at 7.29 to 7.31, it is unlikely to be reasonable to refuse consent because the approval of a superior landlord is required but has not been obtained.

7.42 However, we consider that in some cases it will be reasonable for a landlord to grant consent conditional on the superior landlord's approval.

7.43 The landlord might grant consent but make it conditional on the subtenant obtaining the superior landlord's consent or a declaration that it has been withheld unreasonably if:

- the tenant ought to have applied for the superior landlord's consent but has failed to do so; or

- the superior landlord's consent has been sought and he has not yet responded or he has refused consent on grounds which are not obviously unreasonable.

7.44 It is less likely to be reasonable to grant a conditional consent if:

- the landlord ought to have applied for the superior landlord's consent but has failed to do so; or

- the superior landlord's consent has been sought and he has refused consent on grounds which are obviously unreasonable or has failed to respond for an obviously unreasonable length of time.

In these circumstances, we suggest that the landlord should grant unconditional consent.

Preserving the right to forfeit

7.45 As explained at 7.25 to 7.28, if the landlord is concerned that he may waive the right to forfeit by responding to an application, he may reasonably make consent conditional on the tenant (and any assignee or subtenant) accepting that the consent is without prejudice to any accrued right to forfeit which the landlord might have.

Formal licence

7.46 In many cases, the lease will expressly require a formal licence and reasonableness will be irrelevant.

7.47 Where it does not, we consider that it will often be unreasonable for the landlord to make consent conditional on a formal licence. If he does so, the courts will usually disregard the condition, rather than hold that consent has been granted subject to an unreasonable condition.

Example

A tenant applied for licence for a change from shop use to a restaurant. The landlord's agents wrote to the tenant saying that 'subject to the solicitors drawing up the said deed of variation, your landlord would have no objection to a change of use'. The lease did not state that a deed was required, only written consent. No deed was ever executed.

The landlord was not entitled to demand a deed. Giving consent subject to something that the landlord was not entitled to claim constituted unconditional consent, despite the lack of a deed.[431]

7.48 Sometimes a formal document will be required in order to implement a condition, for example the giving of a guarantee in an assignment case or a covenant to reinstate in an alterations case. In this case, we consider that the landlord can grant consent conditional on completion of a licence giving effect to that condition.

7.49 Sometimes there may be another good reason to insist on a formal licence. If so, we consider that the landlord should make clear what those reasons are when giving consent and imposing the condition of a licence.

7.50 In any case, it is good practice to have the consent clearly recorded, and it may often be appropriate for the parties to enter into a licence after consent has been granted.

Alienation: Reasons for refusal – financial characteristics of the assignee or subtenant

Assignee is a poor covenant

7.51 It is reasonable for a landlord to be concerned about whether the assignee can pay the rent and perform the other covenants in the lease, such as a covenant to repair. If the landlord has reasonable doubts about whether the assignee's financial position is strong enough, it will usually be reasonable to refuse consent.[432] It is not necessary for the landlord also to show that the value of his investment has been diminished.[433]

7.52 Each individual case will turn on its own facts, but factors that the landlord might consider include:

- Can the assignee provide at least three years' accounts? If the assignee could produce accounts but refuses to, it is likely to be reasonable to refuse consent.[434]

- Is the assignee a new company or business?
  - If it has taken over the business of a successful old company as part of a reorganisation, it may be unreasonable to refuse consent.[435]
  - If the new company is a phoenix company that has taken over the business of an insolvent old company, it is likely to be reasonable to refuse consent.
  - If the assignee is starting a new business, it may be appropriate to grant consent only if a suitable guarantee or other security is provided.[436] However, depending on the prospects of the new business, and the amount and nature of the assignee's assets, it may be reasonable to refuse consent altogether or it may be unreasonable to demand security.

- Do the accounts show healthy pre-tax profits? Are profits stable or increasing? If not, it is likely to be reasonable to refuse consent unless there is good reason to expect a turnaround and a suitable guarantee or other security is offered.

- Are low profits or losses in a given year explained by unusual expenditure, such as a re-fit?

- In the case of a small business where there are no employee costs and the assignee takes a personal income from the profit, is the profit before or after directors' or partners' drawings? If before, it may be appropriate to reassess the position after an adjustment has been made. If after, the size of the drawings should be considered. If they are large, will they be reduced if that is necessary to meet obligations?[437]

- Have any recent events changed the assignee's prospects for better or worse? If so, and if the last accounting date was some time ago, are management accounts available to show the up-to-date position?

- Are references favourable? Are they from persons of good standing with sufficient knowledge of the assignee to express an opinion? Do the references contain disclaimers of liability?

- Is the basis of the references clear? For example:
  - Does a trade reference state the level of trade done with the assignee? Is the level of trade sufficiently high to assess creditworthiness?

- Does a landlord's reference state the rent? Is it comparable to the rent of the premises to be assigned? If not, is there other information to demonstrate the assignee's ability to pay a higher rent?

- Has an accountant audited the accounts, or simply relied on figures provided by the assignee? If the accounts are unaudited, would an audit be expected? For example, if a sole trader bases a tax return on unaudited accounts, those accounts may be sufficient.[438]

- Do the accounts and references relate to an existing business which will be transferred to the new property? If the assignee proposes to extend his business or start a new business, the landlord may reasonably want additional comfort.[439]

- Does the assignee have previous business experience?

- Is the assignee able to afford relocation costs without adversely affecting the business?[440]

- Is the assignee a company, or an individual all of whose assets will be available to enforce a court judgment against?

- Are there other sources of income which will help the assignee to meet liabilities under the lease?

- Is stated income before or after tax?[441]

- Are there other liabilities which will affect the assignee's ability to meet obligations under the lease?[442]

- Do assets exceed liabilities? If not, it is likely to be reasonable to refuse consent.

- Does the assignee have large debts repayable on demand? If so, could it pay them? It may be relevant who the creditor is. If the debts are inter-company loans used to purchase fixed assets, and it will not be in the parent company's interest to call the loans in, it may not be reasonable to refuse consent.[443]

- Does the assignee have substantial assets? If so, the landlord may draw comfort, but he is still entitled to be satisfied that the rent will be paid as it falls due.[444] If the assets are cash or other liquid assets, it is less likely to be reasonable to refuse consent; if they are property or other illiquid assets, it is more likely to be reasonable.

- Is the value of the assets clear? For example, is a valuation of property given by a qualified valuer[445] and does it reveal the extent of any borrowing secured against it?[446]

- How long is it before the lease expires? If it is only a short period, the landlord may need stronger doubts to justify a refusal of consent.[447]

- Is the lease onerous? If, for example, the repairing obligations are very light, the landlord may need stronger doubts to justify a refusal of consent.[448]

7.53 Not all of these considerations will be relevant in each case. A good answer to one point may make it unreasonable to give any weight to a bad answer to another point. Landlords should also bear in mind that if they ask for too much information where it is not reasonably necessary, they may be found to have unreasonably withheld consent. See 6.48 to 6.50 for general guidance on asking for further information.

Impact of guarantee

7.54 It is wrong to pay attention only to the covenant of the assignee itself. Instead, the landlord should look at the whole of the situation as it stands. Thus the offer of a guarantee may make it unreasonable to refuse consent to assign even to a covenant of uncertain strength.

Example

A tenant applied for consent to assign a lease to a limited company. The rent payable was £35,000 a year and the rates were £8,000 a year. A premium of £350,000 would be paid (as the market rent was about £100,000 and the next review was not for another three years).

The company's share capital was only £100. The pre-tax profits for the previous year were £13,500 after director's remuneration of £10,000, although the profits had been £56,000 in one of the last three years. The net assets were increasing, but were difficult to value as the company traded in modern art. It had recently borrowed £1m to increase its stock of pictures. It is likely that it would have been reasonable to refuse consent to assign to the company if no guarantor had been offered.

However, the man behind the venture would guarantee all the company's obligations. In addition, the original tenant and its guarantor would both remain liable (as the tenancy was an 'old' tenancy). A refusal of consent was unreasonable.[449]

7.55 Nonetheless, if the tenant's covenant is very poor, the offer of a guarantee may not be enough to make it unreasonable for the landlord to withhold consent. It has been said that 'an assignment to a totally insubstantial company, even though backed by a guarantee, [is] quite a different thing from an assignment to a satisfactory and responsible tenant. It would not be proper to impose such an assignment on the landlord.'[450]

7.56 Further, if a guarantee is offered in principle only, without any indication of its terms, it is less likely to assist the tenant's case.[451]

7.57 If the guarantor is of insufficient quality, it is unlikely to make any difference to the reasonableness of refusing consent.[452]

'New' tenancies v 'old' tenancies

> Most tenancies granted in or after 1996 are 'new' tenancies. Other tenancies are usually called 'old' tenancies. For more details, see Appendix 1.

7.58 If the tenancy is an 'old' tenancy, that is likely to be a relevant factor. This is because there will be other people, in addition to the assignee, who will remain liable for the tenant's obligations until the lease expires. They are:

- the original tenant;
- the original tenant's guarantor; and
- any earlier assignees, if they gave a direct covenant to observe the tenant covenants until the lease expires.

This means that if the proposed assignee defaults in the future, the landlord will be able to look to one or more other people in order to recover the rent or any other losses.

7.59 Conversely, if the tenancy is a 'new' tenancy, the tenant's liability will usually come to an end when he assigns. The same applies to any guarantors.

7.60 In 1986, it was held that the market had little interest in the identity of the original tenant.[453] However, there have been more recent suggestions that the fact that a tenancy is an old tenancy is relevant.[454]

Nonetheless, the fact that an original tenant or some other person with a strong covenant remains liable is not a complete answer to a weak assignee covenant. An original tenant is unlikely to retain any control over the assignee's activities on the premises and its use of them. It may therefore be reasonable for the landlord to refuse consent to assign if the assignee itself will not be a responsible tenant.[455] In this respect, we consider that the continuing liability of a strong original tenant is less valuable to the landlord than a new guarantor; a new guarantor is more likely to be able to exercise some control over the assignee's business and its conduct.

In addition, if the assignee defaults on the rent, the landlord must serve notices and observe time limits (imposed by s. 17 of the LT(C)A 1995) if he wishes to recover the rent from the original tenant. The requirements do not apply if he wishes to recover the rent from a new guarantor. For this reason as well, it is more likely to be reasonable to refuse consent to assign to a weak covenant where there is a strong original tenant who will remain liable than where there is a strong guarantor.

Impact of strong parent company

7.61 In the past, it has been said that 'generally speaking reputable international groups do not allow their subsidiaries to default on their obligations'.[456] This reasoning would make it less reasonable to refuse consent to assign to a weak company if it is supported by a strong parent. However, it is open to doubt whether this principle will survive more difficult financial markets.

Subtenant is a poor covenant: subtenancy outside the LTA 1954

7.62   Assuming that the tenant is a reasonable covenant, the strength of a subtenant is of comparatively little concern to a landlord while the lease lasts: it is the tenant who will be liable to the landlord for the rent and the other covenants. If the subtenant has no statutory protection so the subtenancy will terminate at the same time as, or before, the head-tenancy, it is unlikely to be reasonable to refuse consent to sublet on the basis of the subtenant's covenant strength.

Example

A tenant wished to grant a sublease to a small family company. The landlord refused consent partly on the basis that the company was a poor covenant.

The company had traded in a small way but successfully for six years, four of them in the property opposite the premises concerned from which it needed to relocate. The rent for the premises was £32,500 with the benefit of £12,000 sub-rent from residential accommodation above. The rent for the opposite property was only £14,000. The company's turnover, gross profit and after-tax profits for the previous year were £261,000, £113,000 and £24,000 respectively and for the previous year were £204,000, £89,000 and £14,000. The company was not seeking to expand its business, merely to relocate. Money for fitting out the premises could be provided by the shareholders if necessary.

The court decided that the financial standing of the company could only be a 'contingent concern' of the landlord, and that consent had been withheld unreasonably.[457]

Subtenant is a poor covenant: subtenancy within the LTA 1954

7.63   If the head-tenancy will expire in a few years' time, and the subtenancy will be continued by the LTA 1954, the landlord will have the subtenant foisted upon him as a direct tenant. A weak covenant will therefore be a direct concern to the landlord as this may diminish the capital value of the landlord's investment.

Example

A tenant applied for consent to sublet the whole of its premises. It was common ground that the subtenant's covenant was weak. The lease had about six and a half years left to run, after which the sublease would be continued by the LTA 1954 at a market rent. The subtenant would therefore become the landlord's direct tenant.

Without the subtenant, the building could be re-let in parts when the head-lease expired, producing a higher rent than a single letting. This would also spread the risk of a tenant defaulting. While multiple occupiers would frequently be a disadvantage, it might be preferable to being forced to rely on the covenant of one weak tenant.

Even though this was six and a half years in the future, and it was difficult to predict what would happen, there was evidence that if the freehold was put up for sale immediately after the subletting, a purchaser would reduce his bid to take account of the poorer prospects when the lease expired. On that basis, the landlord's valuer said that the *present* market value of the landlord's investment was £500,000 less with the subtenant in place. It was held to be reasonable to refuse consent to sublet.[458]

7.64 We consider that landlords will only be able to rely on this reason in unusual circumstances. If the only reason why the capital value of the landlord's interest will be reduced is speculation that the subtenant will claim a new lease after the head-lease expires and will be unable to pay the new rent, other cases suggest it may be unreasonable to refuse consent: see 7.70. Legal advice should be sought.

Subtenant is a poor covenant: risk that subtenant will become immediate tenant

7.65 In an appropriate case, it may be possible to take into account the risk that the subtenant might become the landlord's immediate tenant during the term of the existing lease. This might happen if the tenant falls into financial difficulties, and the head-lease is then forfeited or disclaimed. The subtenant will then have the right to claim relief from forfeiture or a vesting order. If granted, the subtenant will become the immediate tenant.

7.66   However, it has been held that this should not be taken into account where the head-lease was for 150 years at a low rent,[459] or where the tenant was a publicly listed company with no history of rent arrears, albeit one with no assets.[460] Even where the tenant is a weak covenant, it may not be reasonable to take this risk into account, as the subtenant's right to become the immediate tenant is discretionary.[461] Further guidance from the courts will be needed and until then, landlords should be cautious about relying on this reason.

Assignee or subtenant does not have pre-tax profits of at least triple the rent

7.67   Pre-tax profits of at least triple the rent may give some basic guidance as to the covenant strength of the proposed tenant. However, this is not a rule to be applied slavishly.[462]

7.68   A tenant with pre-tax profits of ten times the rent for the premises, but which owns a hundred other rented properties and has a large turnover but high and volatile expenses, may be a worse prospect than a stable small business with lower turnover and expenses producing a pre-tax profit of triple the rent. The size of the profits may be of more concern where the assignee or subtenant is expanding its business than when it is merely relocating.[463]

7.69   If the landlord has no other reasons for considering the assignee or subtenant to be a poor covenant, this is likely to be a bad reason.

Assignee or subtenant will have security of tenure and is unlikely to be able to pay the new rent

7.70   This is unlikely to be a reasonable ground for refusal. The courts have held that a landlord is not entitled to take into account the ability of an assignee to pay an increased rent following a renewal under the LTA 1954.

Example

The tenant of a shop wished to assign the remaining two years of the lease. The property was under-rented and it was likely that, if the tenant elected to take a new lease under the LTA 1954, the rent would increase. The landlord argued that it was entitled to consider the proposed assignee's ability to pay that new rent.

This was irrelevant. The tenant might not take a new lease if it could not meet the new rent, and if it proved to be a bad payer the landlord would have grounds to object to the grant of the new lease in any case.[464]

Alienation: Reasons for refusal – other characteristics of the assignee or subtenant

Assignee or subtenant will not run the business successfully

7.71 This may sometimes be a good reason for withholding consent. The success of a business will affect the goodwill of the premises, and this may affect the value of the landlord's interest in the premises. In an appropriate case the landlord is therefore entitled to consider the assignee's experience in running the business in question, the business plan and how profit projections are calculated.

Example

A bank had taken an assignment of a lease of a hotel after the tenant defaulted on its mortgage. It wished to assign the lease on to a newly-formed company with no trading history whose directors had no experience in running hotels. There was no evidence that it intended to engage a hotel management company. The evidence as to the prospects of occupancy levels at the hotel was unclear, nothing like a business plan had been produced, and the landlord was not given any information as to how the profit figure had been calculated.

The court accepted that the success of the hotel would affect goodwill and would reflect on the value of the reversion. Had this reason operated on the landlord's mind and had the assignee failed to offer any explanations, it might have been a good reason to refuse consent.[465]

Assignee or subtenant will gain legal protection

7.72 Many types of lease qualify for some form of legal protection (known generally as security of tenure). An example of such protection is the right of business tenants to a new lease under the LTA 1954.

7.73 Sometimes a tenant will have no security of tenure, but if he assigns or sublets, the assignee or subtenant will have security. This will normally be a reasonable basis for refusal of consent, particularly if granting security is the purpose behind the transaction.[466] Alternatively, the assignment or subletting might change the nature of the security. If this harms the landlord, it is likely to be reasonable to refuse consent.

Example

A business tenant applied for consent to sublet part of his premises. The proposed subtenant would qualify for protection under the LTA 1954.

The landlord refused consent. It said that the tenant might or might not want to take up a fresh lease of the whole premises at the end of the lease. If he did not, the landlord might be forced to grant a new lease to the subtenant of part only.

A refusal of consent on this basis was reasonable.[467]

7.74 We consider that it is unlikely to be reasonable to refuse consent if the present tenant already has security and will simply be transferring his security to the assignee or subtenant.

Assignee or subtenant will gain the right to enfranchise

7.75 Certain tenants have the right to acquire the freehold interest in their property or an extended lease. In the past, it has been held to be reasonable to refuse consent if an assignee would qualify for this right where the tenant does not.[468] However, as almost all tenants now qualify for the right, this will rarely be a relevant reason.

Race, sex, sexual orientation or religion

7.76 A number of statutes make it unlawful for a landlord (or any other person whose consent is required) to discriminate on various grounds by withholding consent to a disposal. In each case, the legislation enables the tenant to bring a claim for damages.[469] It does not expressly state that a discriminatory withholding of consent will be unreasonable, but we consider that it will.

7.77 The grounds are:

- racial grounds, including colour, race, nationality, or ethnic or national origins;[470]
- sex grounds, including sex, gender reassignment, pregnancy or having given birth within the last 26 weeks;[471]
- sexual orientation;[472] and
- religion or belief.[473]

Legal advice should be taken on what constitutes discrimination.

7.78 In the case of race and sex discrimination, these principles apply to assigning, subletting or parting with possession of all tenancies, including agreements for lease, whenever and however they were created.[474] In the case of sexual orientation and religion or belief, they apply to disposal of premises.[475]

Disability

7.79 It is unlawful for a landlord (or any other person whose consent is required) to discriminate by withholding consent to a disposal to a disabled person.[476] Again, the legislation enables the tenant to bring a claim for damages.[477] It does not expressly state that a discriminatory withholding of consent will be unreasonable, but we consider that it will.

7.80 Legal advice should be taken on what constitutes discrimination, and what counts as disability.

7.81 These principles apply to all tenancies, including agreements for lease, whenever and however they were created[478] and to 'disposal' of premises, which is defined as:

- granting a right to occupy premises;
- assigning; and
- subletting or parting with possession of all or part of the premises.[479]

Assignee, subtenant or guarantor registered abroad

7.82 This is not necessarily a good reason for refusing consent.

Example

In August 1992, a tenant sought consent to assign to a newly-formed company with insufficient covenant strength. A guarantor was therefore necessary.

The guarantor offered was a company registered in Luxembourg, mainly concerned with projects outside the UK. It was equally owned by the Kuwaiti Investment Authority and the Algerian Treasury. It had very substantial net assets, including cash in the bank of £5.5m in 1990, £6.5m in 1991 and £7m in 1992, although these assets were not held in the UK. In addition the guarantor offered a rent deposit of a year's rent. It also offered to give an address for service in England, which was to be irrevocable. It was found that judgments could be enforced in Luxembourg with little more difficulty than in England.

In the circumstances, it was not reasonable to withhold consent to assign.[480]

7.83 However, in other circumstances where a foreign registration does adversely affect the landlord, we consider that it might be reasonable to withhold consent.

Diplomatic immunity

7.84 It may be unreasonable to refuse consent to sublet on the basis that the subtenant has diplomatic immunity.

Example

A tenant sought consent to sublet to a counsellor of the Turkish Embassy. The landlord refused consent and argued that, as the subtenant would have diplomatic immunity, he would not be able to sue him for possession if the tenant's lease ended.

The court held that most people would feel certain that the counsellor would comply with the lease, because he was a responsible person and the Turkish government would not allow him to disregard his obligations. Consent had therefore been unreasonably withheld.[481]

7.85   It is unclear whether this decision would be followed in an assignment case, where the landlord has a more direct interest in the lease covenants being legally enforceable against the assignee.

Alienation: Reasons for refusal – the terms of the transaction

Sub-rent is less than market rent

7.86   If the reason for the landlord's objection to the rent is to enhance, or avoid, comparisons which might be detrimental to future rent levels of other shops owned by the landlord, this is an attempt to obtain a collateral advantage. Refusal of consent will be unreasonable.[482]

7.87   However, if the reason for the landlord's objection is that his ability to collect rent, and hence the value of his property, might be adversely affected in the future, a refusal is more likely to be reasonable, at least for commercial premises.

Example

A proposed subletting was to be at a rent well below the current market rent and in consideration of a substantial premium. It was held that the landlord had reasonably withheld consent on the ground that its ability to collect rent, and the value of the property, might be adversely affected in the future.

One concern related to the landlord's rights under the *Law of Distress Amendment Act* 1908. When the head-tenant's rent is in arrear, this Act enables the landlord to require the subtenant to make future payments of rent direct to him. If the rent payable by the subtenant were less than the corresponding rent payable by the head-tenant, the landlord's remedy under this Act would be less effective.[483]

The *Tribunal, Courts and Enforcement Act* 2007 will repeal the *Law of Distress Amendment Act* 1908. It will provide a new basis for collecting rent direct from a subtenant, but this will be more limited. For example, it will only apply to leases of commercial premises. Legal advice should be taken before relying on this reason.

7.88 We consider that this principle is only likely to be relevant where the sub-rent is also lower than the head-rent (or is likely to become lower during the term of the subtenancy: see 7.142 to 7.145).

Sub-rent is less than the head-rent

7.89 Where the sub-rent is a market rent, it is unlikely to be reasonable to withhold consent to a subletting on the ground that the sub-rent is less than the head-rent. To do so would prevent the tenant from subletting at all unless he could find a subtenant to pay an inflated rent, or he was willing to pay a reverse premium.

Example

A tenant had rented a flat at a rent of £700 a year. She wished to join her husband in Peru in a hurry, but could only find a willing subtenant of her flat at an annual rent of £450. The landlords were only willing to agree to this subletting on terms that the subtenant entered into a direct covenant with them to pay the rent of £700 reserved by the lease. It was held that their consent had been unreasonably withheld.[484]

7.90 It might be thought that the landlord could again rely on the fact that his rights under the *Law of Distress Amendment Act* 1908 (or its replacement) will be prejudiced (see 7.86 to 7.88) if the sub-rent is a market rent but less than the head-rent. We do not consider that this will be a good reason to refuse consent: if the tenant is not prepared to pay a reverse premium and cannot find a subtenant to pay the head-rent, the likelihood is that there will be no subtenant at all. This means the landlord's position under the 1908 Act will be even worse.

Reverse premiums

7.91 It follows from the previous section that the fact of a reverse premium being paid will not usually be a good reason to withhold consent. Indeed, in the case of an assignment, a landlord is not usually entitled to know details of any premium or reverse premium being paid.[485]

7.92 If a reverse premium is being paid, that will usually be because the rent is higher than market rent, which is in the landlord's favour. If so, there is unlikely to be any effect on a future rent review, particularly an upwards only rent review.[486] However, in particular circumstances, a reverse premium may adversely affect the landlord, in which case it may be reasonable to withhold consent.

Alienation: Reasons for refusal – the use of the property

Assignee or subtenant will breach the use clause

7.93 If the landlord reasonably considers that the proposed assignment or sublease would or might lead to a breach of the use covenant it will usually be reasonable to refuse consent.

Example

A landlord let a large area of land for development. The lease contained a restriction on use which referred to certain planning use classes.

When the tenant applied for consent to assign the lease, there was a dispute about whether the use restriction applied and whether the assignee would be in breach. The landlord took a reasonable view that the assignee would breach the clause, and refused consent on that basis. The refusal was reasonable.[487]

7.94 We consider that a landlord will not be able to rely on this principle if:

- if the lease allows the use to be changed with landlord's consent not to be unreasonably withheld; and
- the landlord has no good reason to refuse consent to change use.

Assignee or subtenant will use the premises for a permitted purpose which the landlord objects to

7.95 If the landlord reasonably objects to the proposed use, he may sometimes be able reasonably to refuse consent to assign or sublet even if the use is not restricted by the lease.[488]

- Where the lease permits only one type of use, the landlord *cannot* usually withhold consent to assign or sublet on the ground of that proposed use.

Example

A lease contained a covenant allowing only one use: offices with ancillary showrooms. The tenant sought consent to assign to a partnership which proposed to provide serviced office accommodation. It was common ground that this was permitted by the lease.

The landlords refused consent on the ground that 'the investment value of [their] interest in the property would be detrimentally affected by the proposed use'. A refusal of consent meant that the property would be left vacant. The landlord would be fully secured for the rent. The refusal was held to be unreasonable.[489]

- Where the lease restricts only certain types of use, leaving the tenant free to use the premises in any other way, the position is different. An objection to one of those other, permitted, types of use is more likely to be a reasonable ground for refusing consent.[490]

7.96 However, the fact that the lease does not prohibit the proposed use is still a relevant consideration. It may be difficult for the landlord to, in effect, cut down the uses permitted by refusing consent to assign.[491] We consider that the position is likely to depend on the width of the uses permitted by the lease.

Assignee or subtenant will run a competitive trade

7.97 In certain circumstances, this is likely to be a reasonable ground for refusal.

7.98 A landlord can take into account the trading interests of his other tenants, and may refuse consent to assign to a person who would compete with them.[492]

7.99 It is unclear whether a landlord is entitled to take into account his own trading interests. In the past it has been held that it was reasonable to refuse consent to an assignment:

- of next-door premises to a business rival;[493] and

- of a kiosk which would compete with another unit, depressing the rental value of that unit.[494]

7.100 Some doubt has been cast on whether this is right,[495] but we consider that in an appropriate case the landlord can consider his own trading interests.[496]

The transaction will cause traffic difficulties

7.101 It may be that an assignment or subletting will cause increased traffic which the site may be unsuitable for. This is most likely to arise when the tenant applies to sublet part of the premises, if this will increase the number of users of the site. In an appropriate case, this can be a good reason to refuse consent.[497]

Alienation: Reasons for refusal – the transaction will cause harm to the landlord

Difficulty in re-letting other property

7.102 If the proposed assignee is an existing tenant of another of the landlord's properties, the landlord may, by consenting, lose the assignee as his tenant of the other property. Refusal on this basis has been held to be unreasonable, even though the property would be very difficult to re-let.[498]

7.103 This is difficult to reconcile with other cases, and we suggest that legal advice should be taken.

The assignment or subletting will diminish the value of the landlord's interest

7.104 In certain circumstances, this can be a good reason to withhold consent.[499] It is enough that the landlord has genuine and not 'unfounded' concerns on matters relevant to the value of its interest in the property, even if the prospect of those concerns being realised is small.[500] However, in extreme cases the insubstantial nature of the concerns coupled with severe detriment to the tenant if consent is refused may mean that it will be unreasonable to refuse consent. For example, if there is little prospect of the landlord putting the property on the market or mortgaging it to the fullest extent possible, this is less likely to be a good reason.[501]

The diminution will be caused by the weaker covenant of the assignee

7.105 This is unlikely to be a good reason on its own.

7.106 If the present tenant has an extremely strong covenant, such as a government department, almost any assignment will be to a weaker covenant. An assignment to a weaker covenant may well diminish the capital value of the reversion. If this were to be a good reason to withhold consent, then the more substantial the tenant, the more easily a landlord would be able to withhold consent to assign. It has therefore been held that a weaker covenant does not necessarily entitle the landlord to refuse consent.[502]

7.107 Instead, the landlord should weigh the detriment to himself of granting consent against the detriment to the tenant of refusing consent. If the detriment to the tenant of refusing consent is extreme and disproportionate to the benefit to the landlord, it will not be reasonable to refuse.

7.108 Thus it is likely to be reasonable to refuse consent if:

- the covenant of the assignee is objectively weak, and not merely weak by comparison with the existing tenant: see 7.51 to 7.61;
- the weaker covenant will cause a substantial diminution in the value of the landlord's reversion, or the landlord is actively marketing the reversion, at least if that is known to the tenant; or
- there is a good market for the lease, and the tenant is likely to be able to find an assignee with a stronger covenant easily, or to restructure the transaction in a way acceptable to the landlord.[503]

7.109 It is less likely to be reasonable if:

- the covenant of the assignee is perfectly acceptable, but is weak compared to the unusually strong covenant of the existing tenant; or
- the weaker covenant will cause a small diminution in the paper value of the reversion, which the landlord has no intention of realising.[504]

The diminution will be caused by the weak covenant of the subtenant

7.110 This is less likely to diminish the value of the reversion. However, where it does, we consider that similar principles will apply as for an assignment. See 7.63 for an example.

The diminution will be caused by the proposed use

7.111 This can be a good reason in appropriate circumstances, but except in a clear case the landlord will need to have cogent expert evidence to support the assertion that the value will be reduced.[505]

Alienation: Reasons for refusal – pre-existing problems

The premises are in disrepair

7.112 Sometimes the tenant will be in breach of a covenant to repair the premises. This is only sometimes a good ground for refusal. It depends very much on the seriousness of the disrepair and the ability and willingness of the assignee or subtenant to remedy the problems. The following principles apply:

- The starting point is that the mere fact that the premises are in disrepair, at least where the problems are not very serious, does not entitle the landlord to refuse consent.[506]

- However, it is not in general unreasonable of a landlord to refuse its consent if:
  - there are breaches of the covenant to repair that are more than minimal, and more especially if they are extensive and longstanding; and
  - the landlord cannot be reasonably satisfied that the proposed assignee or subtenant will remedy them.[507]

  The assignee's ability to fund the repairs may be an important consideration.

- The more serious the tenant's default, the more likely it is that the landlord can reasonably refuse consent.[508]

Examples

(1) A tenant held a lease of a warehouse and office unit, at an annual rent of £400,000. The property had become dilapidated, and the landlord had served the tenant with notice to repair. It estimated the cost of repairs to be around £500,000.

The tenant then applied for consent to sublet, on the basis that the subtenant would not take on the repairing covenant and would only be required to keep the property in its present condition. The landlord refused consent unless the issue of the disrepair was resolved.

The court held that the dilapidations were extensive and longstanding, and that the terms of the sublease showed that the subtenant would not carry out the repairs itself. Although the court decided that, for other reasons, the application had not been valid, consent could have been reasonably refused on the basis of the disrepair.[509]

(2) A tenant held a long lease of a terrace in central London, which it decided to put up for sale. A few years earlier the landlord had obtained a surveyor's report showing no serious defects. However, after learning that the lease was for sale (and hoping to acquire it), he hastily arranged a further report.

When the tenant had found a buyer and applied for consent to assign, the landlord served a schedule of dilapidations and refused consent. The tenant sued, and the court found that the defects were no more serious than those in the earlier report. The disrepair was neither extensive nor longstanding, nor unlikely to be remedied by the assignee. The refusal was unreasonable.[510]

The tenant is in breach of other covenants

7.113 If the breaches are minor, or are unlikely to continue and are remediable, it is unlikely to be reasonable to refuse consent.

Example

A tenant applied for consent to sublet a retail shop. The landlord refused consent because of several breaches of the lease. These were all categorised by the landlord (and later by the court) as 'minor'. They included the fact that the proposed subtenant had been allowed into occupation before the tenant had applied for consent, objections to the tenant's signage, and fees owed for a previous abortive application for consent.

The court decided that these were not valid reasons to refuse consent. The breaches were all minor, were unlikely to continue and were remediable.[511]

7.114 If the breaches are more serious, we consider that the principles will be similar to those for failure to repair.

Alienation: Reasons for refusal – the landlord's plans for the property

Landlord wants to recover possession for himself

7.115 Unless the lease contains very unusual provisions, this will almost always be an advantage which the landlord has not covenanted for. This will therefore be a bad reason.

7.116 Landlords will rarely expressly give this as a reason. However, the course of the landlord's conduct may make clear that this is the real reason.

- The landlord may have given no good reason, in which case consent will of course have been withheld unreasonably.
- Alternatively, the landlord may have given a good reason as well. If so:
  - in many cases the landlord's real motive will not mean that consent has been unreasonably withheld; but
  - in some cases the good reason will clearly be a makeweight, or the landlord's motive will vitiate the good reason. In those cases, consent may still be found to have been unreasonably withheld.

Example

A tenant applied for consent to assign the lease of a shop. The landlord made numerous detailed enquiries about the proposed assignee and her financial status, and refused to make a decision. This ultimately led the assignee to withdraw from the transaction.

The court held that the landlord had, in fact, been motivated by a desire to stop the assignment and take a surrender of the lease. Consent had been unreasonably withheld.[512]

Good estate management

7.117 Used properly, this may be a good reason to refuse consent to a transaction. For example, if the landlord's policy is to promote the creation of larger units, then depending on the terms of the lease it may be reasonable to refuse consent to a subletting of part of a unit only.[513]

However, the policy must be rational, and cannot be used simply as a blanket ban on certain transactions.

Examples

(1) A tenant, a men's fashion retailer, had a lease of a shopping centre unit. The lease restricted the use of the premises to the sale of clothing, shoes and so on, and such other items as the landlord might approve. That approval was not to be unreasonably withheld, provided it was consistent with good estate management and the distribution of trades in the shopping centre.

The tenant sought consent to assign. As the proposed assignee was an electronic games company, the tenant also applied for consent to change of use.

The landlord refused consent on the basis that the proposed assignee did not conform to its tenant mix policy. The landlord had a policy of reserving the area around the unit for fashion retailers, which the court held was rational and could be enforced. The refusal of consent was reasonable.[514]

(2) A landlord withheld consent to an assignment because they had a policy of refusing all such applications. They had a waiting list of potential tenants and did not wish to allow tenants to choose their successors. The refusal was unreasonable.[515]

Alienation: Imposing conditions – additional security for the landlord

Provision of a direct covenant by the assignee or subtenant to observe the tenant covenants

7.118 Depending on the period for which the covenant will last, we consider that this will usually be a reasonable condition. Subtenants should of course not be required to covenant to pay the head-rent, or to observe any other covenants which should be performed by the head-tenant.

7.119 Frequently, this is not a reasonableness issue. Many 'old' tenancies contain an express requirement that any assignee will covenant directly with the landlord to observe the tenant covenants for the whole of the remainder of the term. As explained at 1.36 to 1.40, if this express requirement is a pre-condition, the landlord will be able to insist on it.

This improves the landlord's position. Without it, the assignee will cease to be liable to the landlord if he in turn assigns the tenancy on. If, on the other hand, he gives a direct covenant, the landlord will be able to look to the assignee as well as the original tenant if, for example, a subsequent assignee defaults on the rent.

Most tenancies granted in or after 1996 are 'new' tenancies. Other tenancies are often called 'old' tenancies. See Appendix 1 for more details.

7.120 If there is no pre-condition, we consider that the landlord will be able to insist on the assignee giving a direct covenant, but only one limited until the next lawful assignment (i.e. one which takes place with the landlord's consent or otherwise in accordance with the lease). The reasons for this are that:

- A covenant which lasts beyond the next lawful assignment will increase or enhance the landlord's rights under the lease, and will therefore be unreasonable.[516] In the case of a 'new' tenancy, the law also renders such a covenant void, even if it is required by a pre-condition or an agreement on reasonableness.[517]

- Conversely, a covenant which lasts only until the next assignment (lawful or not) may offer insufficient protection for the landlord. This is because the assignee would be able to terminate its obligations by assigning in breach of covenant to an insubstantial company.

7.121 We therefore think that it will be reasonable for a landlord to insist that the covenant should last until the next *lawful* assignment. Landlords should consider taking this direct covenant even in the case of a new tenancy; if this covenant is not taken, it is probable that the assignee will be released on an unlawful assignment, although the law is not clear.

Provision of a new guarantor

7.122 If the landlord could not refuse consent outright on the basis of the financial standing of the assignee or subtenant, it will rarely (if ever) be reasonable to require a guarantor. But if consent could be refused, it will frequently be reasonable to grant consent conditional on provision of a suitable guarantee instead. Guarantees will frequently be given by directors of companies, or by parent companies.

Example

A tenant applied for consent to assign a shop to a wholly owned subsidiary of a FTSE 100 company. Another subsidiary, a property holding company, was offered as guarantor. The landlord demanded a guarantee from the parent company.

The assignee was a new company with no track record. It was therefore reasonable for the landlord to require a guarantee. The question was whether the property holding subsidiary was a suitable guarantor.

The property holding company had share capital of only £2, but in the previous year it had made an operating profit of almost 100 times the rent for the shop, it had net assets of over £13.5m and its parent company was of high standing. The court held that there was no rational basis for assuming that the parent might allow the subsidiary to default on its obligations. The demand for a guarantee from the parent was unreasonable.[518]

7.123 Leases frequently contain pre-conditions or agreements on reasonableness which require a guarantee to be given. If so, then as usual the landlord will be able to demand a guarantee even if it is not reasonable. See 7.1 to 7.13.

7.124 Special rules apply to guarantees given by outgoing tenants on an assignment; these are explained at 7.130 to 7.137.

Nature of the property

7.125 In some cases the security of the property itself may make a guarantee unnecessary and so unreasonable. In other cases, the security of the property may be an inconvenient alternative to a guarantee, in which case it is more likely to be reasonable for the landlord to require a guarantee.

Examples

(1) A tenant may wish to assign a long residential lease granted for a premium with a low rent to a company with a worthless covenant. The company has an interest in paying the rent and observing the covenants to preserve the valuable lease. If the company does default, the landlord can forfeit the lease and re-let it for a further premium. It is unlikely to be reasonable for the landlord to require a guarantee.

(2) A block of flats is let on long leases to a large number of subtenants. A substantial rent is payable under the head-lease. The head-tenant wishes to assign to a company with a poor covenant.

The landlord is a retired individual. If the assignee defaults, and the landlord forfeits the head-lease, he is likely to be faced with numerous applications for relief from forfeiture from the subtenants. He may then have to manage the property himself, providing services and collecting service charge.

As he is unwilling to engage in the business of property management, this will be an inappropriate alternative to a straightforward guarantee. It is likely to be reasonable for him to require a guarantee as a condition of granting consent to assign.

(3) Another block of flats is let on short leases at rack rents to a large number of assured shorthold subtenants. The head-lease, with the right to receive rents substantially higher than the head-rent, is valuable.

The assignee has an interest in paying the head-rent and observing the covenants, to preserve the valuable head-lease. It may therefore be thought that this is like Example 1. However, the assignee might terminate the subtenancies and grant new long leases at premiums. This will reduce the value of the head-lease. The position may then be more like Example 2. It is therefore likely to be reasonable for the landlord to require a guarantee as a condition of granting consent to assign.[519]

Duration of a guarantee: next lawful assignment

7.126 As with a direct covenant, it will not usually be reasonable to insist on a guarantee which will last beyond the next lawful assignment by the assignee.[520] We consider that it will be reasonable to link the automatic release of a guarantee to a *lawful* assignment, and not an unlawful one.[521] Compare 7.118 to 7.121.

7.127 If the landlord wants to impose a condition on the release, he should be careful how he words it.

- The landlord may demand a guarantee which will be released on the next lawful assignment as long as 'reasonable alternative security' is provided. A future assignee with a strong covenant may itself constitute reasonable alternative security. This formulation therefore adds nothing to the landlord's right to refuse consent to or impose conditions on that further assignment and so is reasonable.[522]

- If, however, the landlord demands a guarantee which will be released on the next lawful assignment as long as a 'reasonable alternative guarantor' is provided, this is likely to be unreasonable. If the assignee wished to assign on to a very strong covenant, the landlord would not normally be able to demand a guarantee on that further assignment. However, this formulation would allow him to require the existing guarantor to continue to be liable unless the subsequent assignee did provide a guarantor.[523]

Duration of a guarantee: three years' accounts

7.128 It is common for three years' accounts to be provided as evidence of an assignee's covenant strength. As an extension of this, it is common for a guarantee to be released as soon as the assignee has produced satisfactory accounts for three consecutive years, and in many cases it may be unreasonable for a landlord to require a longer lasting guarantee.

7.129 However, while this principle works well for trading tenants whose business provides the income from which the rent will be paid, it is less appropriate for investment tenants who will pay the rent from the income from subleases. This is particularly so for residential subleases, where new leases may be granted for a premium and a peppercorn rent. Satisfactory accounts for three years by a trader indicate that the trader is making a success of his business, which may reasonably be expected to continue. Satisfactory accounts from a property investor in relation to the property in question indicate only that the premises are presently rent-producing; there is no guidance as to the position for the future.[524]

Provision of a guarantee by the existing or previous tenant

7.130 The position here depends on whether the tenancy is an 'old' tenancy or a 'new' tenancy.

> Most tenancies granted in or after 1996 are 'new' tenancies. Other tenancies are often called 'old' tenancies. See Appendix 1 for more details.

'Old' tenancies

7.131 For 'old' tenancies, a guarantee from the outgoing tenant or any previous tenant will rarely be needed. The original tenant will remain liable anyway, as will any guarantor of his. An assignee and any other guarantor who has given a direct covenant to comply with the tenant covenants for the whole of the term will also remain liable anyway.

7.132 Where the outgoing tenant will not be liable anyway (because he was not the original tenant and has not given a direct covenant), we consider that the position will be the same as that discussed at 7.122 to 7.129 for other guarantees. In addition, it is unlikely to be reasonable to require the outgoing tenant to stand as guarantor for the incoming tenant if the incoming tenant is offering a suitable new guarantor of his own.

'New' tenancies

7.133 For new tenancies, the original tenant will usually be released from the tenant covenants unless the assignment is in breach of covenant or is an assignment 'by operation of law'.[525] The same applies to any later tenant who has given a direct covenant to comply with the tenant covenants. It is probable that a later tenant who has not given a direct covenant will be released in any event (i.e. whether or not the assignment is lawful), but the law is not clear.

7.134 The tenant can be required to act as guarantor for the assignee, but the guarantee must comply with special requirements.[526] This type of guarantee is called an authorised guarantee agreement (AGA).

7.135 If the tenant who wants to assign ('T2') became the tenant as the result of an assignment from a previous tenant ('T1') in breach of covenant or by operation of law, T1 can be required to enter into an AGA as well as T2.[527] This may not apply if T1 was not the original tenant and did not give a direct covenant to the landlord to comply with the tenant covenants for the whole of the term. In this case, legal advice should be taken.

7.136 Legal advice should be taken as to the precise requirements, but the most important are:

- An AGA can only be demanded where:
  - the covenant is absolute, or is qualified in one of the rare cases where it is not made fully qualified; or
  - there is a pre-condition (for example, *'not to assign without entering into an AGA'*); or
  - the covenant is fully qualified, and there is an agreement on reasonableness which requires the outgoing tenant to enter into an AGA; or
  - the covenant is fully qualified, and it is reasonable to demand an AGA.[528] We consider that it will be reasonable to demand an AGA in the same circumstances as it is reasonable to require the outgoing tenant of an old lease who would not remain liable anyway to provide a guarantee.

See 1.6 to 1.15 for information on when covenants are absolute, qualified or fully qualified, and 1.33 to 1.49 for information about pre-conditions and agreements on reasonableness.

See Appendix 2 for details of when qualified covenants are not made fully qualified.

- An AGA may not:
  - require the outgoing tenant to guarantee the obligations of anyone except the incoming tenant; or
  - impose any liability on the outgoing tenant in relation to any time after the incoming tenant is released from the tenant covenants.

  If it does, the offending requirements will be ignored.

7.137 Any guarantor of the outgoing tenant will also be released on the assignment, unless it is in breach of covenant or is an assignment by operation of law. It is unclear whether that guarantor can be required to guarantee an outgoing tenant's AGA.

Provision of a rent deposit

7.138 If the landlord could not refuse consent outright on the basis of the financial standing of the assignee or subtenant, it will rarely (if ever) be reasonable to require a rent deposit. If consent could be refused, it will frequently be reasonable to grant consent conditional on provision of a suitable rent deposit instead.

Alienation: Imposing conditions – the terms of the transaction

Subtenancy to be 'contracted out' of the protection of the LTA 1954

7.139 In certain circumstances, it may be reasonable to withhold consent to sublet on the basis that a subtenant will have the right to a renewal lease under the LTA 1954. The landlord's reason may be that:

- the tenant himself would not have had that protection (see 7.72 to 7.75); or
- the prospect of the subtenant taking a renewal lease causes damage to the value of the landlord's property (see 7.63 to 7.64).

7.140 If consent could reasonably be withheld on this basis, it is likely to be reasonable instead to grant consent subject to a condition that the subtenancy is 'contracted out'. This is a process whereby a tenant gives up the rights he would have had under the LTA 1954.

7.141 Where it is reasonable to insist that a subtenancy is contracted out, we also consider that it will be reasonable for the landlord to insist on seeing copies of the relevant documentation, so that he can check that the contracting out has been validly effected.

Subtenancy to be on the same terms as the head-lease, or on other specified reasonable terms

7.142 We consider that this is likely to be a reasonable condition in most cases where there is a reasonable prospect that the subtenant will become the landlord's direct tenant.

7.143 For example, if the whole of business premises are sublet on a lease which is not 'contracted out' of the LTA 1954, the head-tenancy is likely to end on its expiry date while the subtenancy will be continued by the Act. This will mean that the subtenant becomes the landlord's direct tenant on the terms of the subtenancy. If the subtenant applies for a new tenancy, it is likely to be on the same or similar terms to the previous subtenancy. The landlord therefore has a direct interest in the terms of the subtenancy.[529]

7.144 However, we consider that the landlord will not be able to insist on unreasonable terms. For example, if the landlord seeks to insist on an absolute prohibition on assigning or subletting the sublease, contrary to the recommendations of the Code for Leasing Business Premises in England and Wales 2007, this may not be a reasonable condition.

7.145 If there is no likelihood of the subtenant becoming the landlord's direct tenant, it is less likely to be reasonable to impose a condition relating to the terms of the sublease, unless the landlord has some other direct interest in the terms of the sublease.

Examples

(1) A tailor owned a property and occupied part of it. The remainder was let on a lease which prohibited certain offensive trades such as brewer or butcher. The tenant applied for consent to underlet to someone who would use part as a showroom and sub-underlet the remainder when a suitable sub-undertenant could

be found. The landlord granted consent subject to a condition that the undertenant agreed not to assign or sublet without his consent, not to be unreasonably withheld.

The court held that the condition was reasonable.[530] We consider that the same is likely to apply even where the landlord does not occupy neighbouring premises.

(2) A commercial tenant has a lease at a rent substantially below the market rent. He wishes to sublet at the passing rent, and take a premium to reflect the low rent. Unless expressly prohibited by the lease, this is likely to be reasonable: see 7.86 to 7.88.

The head-rent is due to be reviewed in two years time. If the sub-rent is not reviewed on similar terms at the same time, the landlord's ability to collect the rent if the head-tenant defaults may be prejudiced (again, see 7.86 to 7.88). It is therefore likely to be reasonable for the landlord to make consent to sublet conditional on the subtenancy containing a similar rent review clause to the head-tenancy.

Alterations: Reasons for refusal

The value of the landlord's interest in the premises or any neighbouring premises will be diminished

7.146 This is a bad reason to refuse consent. Instead the landlord should grant consent, subject to payment of appropriate compensation.[531] See 7.152 to 7.153.

Structural effect of the alterations

7.147 A reasonable concern about the structural effect of alterations is a good ground for refusal. The landlord is entitled to know the tenant's solution to any structural problem, and need not simply give consent conditional on the problem being resolved.

Example

A tenant applied for consent to convert the ground floor of a building into a restaurant. The landlords had reasonable concerns about the impact this would have on the structure of the remainder of the building, which was being converted into flats. The court held that refusal on that basis was reasonable.[532]

The appearance of the alterations

7.148 The landlord may reasonably refuse consent on aesthetic, historic or sentimental grounds.[533]

Examples

(1) A tenant sought consent for wide-ranging improvements to shop premises. The court held that visual and architectural considerations could be good grounds for refusing consent. In that case, however, the works were to a 'confused area of garages, yards and sheds' to the rear of the property and no such considerations arose.[534]

(2) The trustees of a residential estate refused consent to pave a front garden and remove a boundary wall. The decision was based on the impact the works would have on the appearance of the area. The court decided that the refusal was reasonable, notwithstanding that a number of other gardens in the area had been paved.[535]

Use of the altered premises

7.149 An objection to the use that will be made of the premises following the alterations may be a good ground for refusal of consent. This is so even if the new use is not prohibited by the lease.[536]

7.150 Whether a refusal on this basis will, in fact, be reasonable depends on the precise circumstances of the case. If the new use is permitted by the lease, and the tenant's intention to convert the property to that use was known to the landlord when he acquired the property, it might be unreasonable to refuse.[537]

Good estate management

7.151 Good estate management is more likely to be relevant to applications to change use. But in certain circumstances, it can provide a good reason to withhold consent to an application to make alterations.

Example

The Crown Estate Commissioners had a policy of strengthening Regent Street as both a shopping street and a business location and improving the Public Realm. The main objectives included new, larger and better shops. In conjunction with objections to a change of use, discussed at 7.172, it was reasonable to object to an application to subdivide one unit into two, as this conflicted with the policy of encouraging an increase rather than a decrease in the extent of shop frontage to Regent Street and the size of individual units.[538]

Alterations: Imposing conditions

Payment of compensation

7.152 The landlord may require payment of reasonable compensation in respect of any damage to the value of the premises or any of his neighbouring premises.[539] In fact, if the loss of value is the landlord's only objection, he cannot refuse consent outright: this condition is his only option.[540]

7.153 For further details, including how the landlord should phrase a request for compensation and the position where there is an RTM company, see 3.45 to 3.50.

Covenant to reinstate

7.154 Whether it is reasonable to demand a covenant to reinstate depends on the background circumstances and the terms of the proposed obligation.

7.155 The law specifically permits the landlord to require that the premises are reinstated to their prior condition if:

- the alteration 'does not add to the letting value of the holding'; and
- 'such a requirement would be reasonable'.[541]

Strictly, this applies only to 'improvements' but we consider it likely that the same principle would apply to other alterations.

7.156 The fact that the requirement must be reasonable renders this rule somewhat circular, since the landlord must act reasonably in any case. However, we consider that it will usually be reasonable to require reinstatement If the first point is satisfied, unless there is a good reason to the contrary. This might be, for example, that the landlord will definitely demolish the premises after the end of the lease.

7.157 It is not clear whether this rule is intended to set out the *only* circumstances in which a reinstatement obligation can be imposed. For example, if the alteration *does* add to the letting value of the holding, could it still be reasonable to require reinstatement?

Compliance with regulations

7.158 It is likely to be reasonable to require the tenant to obtain building regulation approval and planning permission for the works. A failure to comply is likely, in either case, to involve enforcement proceedings. This is a clear detriment to the landlord which he is entitled to avoid.

7.159 We therefore consider that it is both reasonable and wise for a landlord to impose conditions that the tenant:

- observes all relevant building and health and safety regulations and obtains the relevant approvals;
- applies for all necessary planning consents;
- complies with all planning conditions imposed; and
- does not begin the works until planning permission has been obtained.

Supervision of works

7.160 It is unclear whether a landlord can reasonably insist that he supervises the works as a condition of consent.

7.161 We regard it as unlikely to be reasonable where the tenant is engaging competent contractors to do the works and will covenant to complete them in a workmanlike manner, particularly where they do not affect the structure of the building. The landlord will have had the opportunity to satisfy himself that the works themselves are suitable (see 7.147 to 7.148).

Other conditions

7.162 Other conditions are likely to be reasonable insofar as they are directed at curing a problem to which the landlord can reasonably object.

7.163 They should, however, go no further than is necessary to deal with the problem. Provided that is the case, the landlord is likely to be on safer ground by consenting subject to a carefully considered condition, than by refusing consent outright.

Example

A tenant had a lease of a hotel and conference centre. The landlords owned a nearby property, which they used for weddings and functions. The tenant applied for consent to alter and extend the premises.

The landlords were concerned that the tenant would use the extended property to compete with their wedding business. They therefore gave consent, but attached a condition that the extension could not be used for anything other than management training conferences.

The court decided that it was reasonable of the landlords to want to protect their business from competition. This would not have justified outright refusal, but it did justify imposing a condition. However, by restricting the tenant to management training conferences only and by not qualifying the restriction to allow other uses with landlord's consent not to be unreasonably withheld, the condition went further than was necessary to protect the landlords and was unreasonable.[542]

Change of use: Reasons for refusal – harm to the landlord

The value of the landlord's interest in the premises or any neighbouring premises will be diminished

7.164 In some cases, the proposed use might reduce the letting value of the premises (or nearby premises owned by the landlord) on a lease renewal or a rent review. The capital value of the landlord's interests might be reduced as well or instead. It is likely that this will be a good reason to refuse consent[543] but the position is not clear.

Use would adversely affect the area

7.165 An adverse impact on the surrounding area can be a good reason to withhold consent, at least if the surrounding properties are owned by the same landlord.

7.166 The same may apply even where the landlord does not own other property. If the new use is likely to harm the neighbourhood, rents are likely to be driven down. This could harm the landlord's interest on a lease renewal, or even a rent review.

Example

A tenant applied to change use from a restaurant to what was described as a restaurant with amusement facilities, but was more accurately an amusement hall with ancillary refreshments provided.

The landlord (a city council) refused consent on the basis that the new use 'could and probably would attract an undesirable element to the area' leading to 'vandalism and disorder generally' and 'the true middle aged and respectable shopper tending to shun not only the premises, but the immediate area surrounding them. This could in turn lead to an undermining of the economic fabric of the immediate area.' The reason was held to be good.[544]

Use will compete with landlord's business

7.167 A reasonable concern that the new use will compete with the landlord's own business is likely to be a good ground for refusal.

Example

The tenant of a squash club and gym applied for consent to change the use to a health club with swimming pool. The landlord (a local council) ran a swimming pool next door and a sports centre close by.

The landlord refused consent on the basis that the new use would compete with its own business. The court held that this was reasonable.[545]

7.168 If the tenant has no other realistic use for the premises (so that a refusal of consent would effectively sterilise them for the remainder of the lease) this may cease to be a good ground. In such cases, the landlord must weigh the consequences for its own business against the effect of refusal on the tenant.[546]

Loss of planning permission

7.169 If the change of use will mean that the ability to use the premises for their current purpose will be lost under planning laws and this will have an adverse effect on the landlord, this may be a good reason.

Example

A unit with no street frontage was used as a reception, consulting room and offices for a medical practice. The tenant applied for consent to assign. The assignee was already the tenant of the neighbouring unit, and proposed to use both units together as a travel agency, a retail use. The lease of the neighbouring unit expired many years later than the lease of the first unit.

If the planning permission to change the first unit to retail use was implemented, the existing use would be lost. There was no guarantee that a further planning application, either to reinstate the previous use or to use as offices, would be successful. The first unit was unmarketable for retail user unless it was used with the neighbouring unit. The landlord was held to have acted reasonably in refusing consent to change use.[547]

7.170 It will usually be necessary for the landlord to produce some evidence suggesting that the planning position would be adversely affected. In the example given above, it was sufficient that the assignee had applied for planning permission for use as offices (which would have protected the landlord's position), but the council had required the application to be amended to seek permission for retail use.

7.171 However, in other cases, this may be a bad reason.

Example

A lease contains a covenant 'not to use the demised premises except as a bank within class A2, or for such other use within classes A1 or A2 as may first be approved by the landlord, such approval not to be

---

209

unreasonably withheld'. Given the express contemplation that the tenant might change the use to an A1 use, it is unlikely to be reasonable to withhold consent to such a change on the basis that A2 use might be permanently lost.

Change of use: Reasons for refusal – the landlord's plans for the property

Good estate management

7.172 Used properly, this can be a good reason to refuse consent. Where the landlord owns an identifiable coherent estate and has a strategy for preserving or enhancing that estate, it may be reasonable to refuse consent to a change of use which is not consistent with that policy. The policy need not have been in existence at the beginning of the lease.

> *Example*
>
> The Crown Estate Commissioners had a policy of strengthening Regent Street as both a shopping street and a business location and improving the Public Realm. The main objectives included: maintenance of the strong retail frontage at ground floor level; new, larger and better shops, especially on the west side; and creation of some specific 'magnet' stores. It was reasonable to refuse consent to a change of use from retail to part service use, as this conflicted with a reasonable, known, estate management policy.[548]

Tenant mix

7.173 Good estate management may include securing a good tenant mix. A landlord is entitled to object to a change of use if the proposed new use is already well-represented in the landlord's estate and the change of use will detract from the tenant mix.[549]

Style of frontage

7.174 Good estate management may also include considerations of trading style. Particular uses may have styles which are inconsistent with the landlord's estate management policy.

Example

A fruit and vegetable retailer found his business had become uneconomic when a supermarket opened nearby. It was difficult to find another retailer to take the lease. He applied for consent to assign to a building society. The landlord refused consent on the basis that financial institutions had 'dead' frontages and there were already many financial institutions in the shopping centre. Its decision was found to be reasonable.[550]

Increase in traffic

7.175 It may be that a different use will cause the amount of traffic to increase or the type of traffic to change, which the site may not be suitable for. In an appropriate case, we consider that this would be a good reason to refuse consent.[551]

Limitations of 'good estate management'

7.176 Using the mantra of 'good estate management' will not allow the landlord to secure a collateral advantage: see 7.21 and 7.178. Landlords should therefore use this reason with care, and should check that the policy is consistent with the lease.

7.177 Similarly, the landlord cannot reasonably rely on good estate management where the basis for the refusal is factually incorrect.

Example

A landlord refused consent to sublet and to change use from precision engineering to print finishing and packing. The reason given was that the application to change use conflicted with its interpretation of good estate management in that it would lead to an over-intensification of the use of the site, causing traffic congestion. In fact there would be no traffic problems that could not be overcome easily. The refusal was unreasonable.[552]

Landlord wants a particular new use

7.178 This will usually be a bad reason as it will usually amount to a collateral advantage. Landlords should consider each application on its merits. If a particular use is desired, the landlord should have ensured that the use clause in the lease reflected that desire.

Example

A lease contained a covenant '*not without the consent in writing of the landlord to carry on any trade or business other than as a travel and employment bureau and theatre ticket agency except that the use may be changed with the consent of the landlord such consent not to be unreasonably withheld*'.

A building society wished to acquire the lease. An application was made to change the use accordingly. The local authority landlord refused, on the basis that service occupancy had a depressing effect on neighbouring rents compared to retail occupancy, and it had a duty to ratepayers to maximise rents. Use as a building society was, however, no different from use as an employment bureau (the existing actual use).

Achieving a retail tenant would be a collateral advantage for the landlord, wholly unconnected with the terms of the lease. A refusal designed to achieve that advantage was therefore unreasonable.[553]

7.179 However, the position would be different if the use that the landlord wants is the use that is specified in the lease.

Example

If, in the previous example, the existing use clause had been for one or more retail uses, the landlord would probably have been reasonable in refusing consent for use as a building society in order to retain retail use. See 7.23.

Change of use: Imposing conditions

Payment of compensation

7.180 In the case of qualified covenants, a landlord may require payment of 'a reasonable sum in respect of any damage to or diminution in the value of the premises or any neighbouring premises belonging to him'. If there is disagreement about what the reasonable sum is, the court can determine it, and the landlord will then be bound to grant the licence on payment of that sum:[554] see 4.53 to 4.60.

7.181 We consider that the same will apply to fully qualified covenants, but the position is not clear.

Obtain planning permission and comply with planning conditions

7.182 Many changes of use will require a new planning permission. If the tenant implements a new use without obtaining planning permission, both the tenant and the landlord risk having an enforcement notice served on them.

7.183 We therefore consider that (similar to the position on alterations) it is both reasonable and wise for a landlord to impose a condition that the tenant applies for all necessary planning consents, complies with all planning conditions imposed, and does not implement the change of use until planning permission has been obtained.

Other conditions

7.184 Each case must be considered on its own facts. If the proposed use will pose a potential problem which can be overcome by a condition, landlords should consider granting consent subject to the condition rather than refusing outright.

Example

A tenant of retail premises with residential accommodation above wishes to grant a sublease of the retail premises to the owner of a pet shop.

The landlord may have concerns that the new business will include noisy pets and adversely affect his ability to sublet the residential premises. He could grant consent subject to a condition that no noisy animals are to be kept on the premises.[555]

Special cases

Secure tenancies: exchanges

> A secure tenancy is a type of residential public sector tenancy. For further details, see Appendix 1.

7.185 If a secure tenant applies to exchange his tenancy with another tenant (see 2.41), the landlord may only refuse consent on specified grounds. Those grounds are set out in Appendix 3 (at pp. 296 to 299).[556] If the landlord refuses consent on any other grounds, consent is deemed to have been given.

7.186 The only condition which may be put on a consent is that any outstanding rent must be paid, any breach of covenant must be remedied, and any outstanding obligation under the lease must be performed. If the landlord attempts to impose any other condition, it will be disregarded.[557]

Secure tenancies: subletting part

7.187 In addition to the considerations which apply to all tenancies, the law expressly provides that the landlord can rely on the following considerations:

- the consent would lead to overcrowding of the dwelling-house;[558] and
- the landlord proposes to carry out works on the dwelling-house, or on the building of which it forms part, and the proposed works will affect the accommodation likely to be used by the subtenant who would reside in the dwelling-house as a result of the consent.

The landlord must show that the considerations apply to the particular application.

7.188 The landlord may not impose any conditions. If it attempts to do so, it will be treated as having given consent unconditionally.[559]

Alterations to secure, protected and statutory tenancies

A secure tenancy is a type of public sector residential tenancy. Protected and statutory tenancies are types of private sector residential tenancy usually granted before 15 January 1989. For further details, see Appendix 1.

7.189 Secure, protected and statutory tenancies are subject to an implied term that alterations or additions cannot be carried out without landlord's consent, which is not to be unreasonably withheld. This is explained at 3.31 to 3.34.

7.190 The law states that the following factors are of particular relevance to the reasonableness of a refusal:

- whether the works are likely to make the property, or any other property, less safe for occupiers;

- whether the works are likely to cause the landlord to incur extra expenditure; and

- whether the works are likely to reduce the selling or letting value of the property.[560]

The reasons and conditions explained at 7.146 to 7.163 may also be relevant.

Alterations to comply with disability laws

7.191 As explained at 3.56 to 3.60, restrictions on alterations are sometimes modified to allow the tenant to comply with disability laws.

7.192 Where the regime applies, it imposes rules about how to judge the reasonableness of the landlord's response. The rules are not drafted in plain English and we suggest that legal advice is taken in all but the simplest cases. The rules which apply where an employer needs to make reasonable adjustments to premises are:

- The landlord will be taken to have withheld consent unreasonably where:
  - the lease provides that consent shall or will be given to an alteration of the kind in question and he withholds his consent to the alteration; or
  - the lease provides that consent shall or will be given to an alteration of the kind in question if it

is sought in a particular way, it has been sought in that way and he withholds his consent to the alteration, or

- the landlord has failed to respond within set time limits (explained at 6.36 to 6.41).

- The landlord will be taken to have withheld consent reasonably where:
  - there is a binding obligation requiring the consent of any person to the alteration, the landlord has taken steps to seek that consent, and that consent has not been given or has been given subject to a condition making it reasonable for him to withhold his consent; or
  - the landlord is bound by an agreement which allows him to consent to the alteration in question subject to a condition that he makes a payment, and that condition does not permit the landlord to make his own consent subject to a condition that the occupier reimburse him the payment.

- The landlord may reasonably impose conditions to the effect that:
  - the occupier must obtain any necessary planning permission and any other consent or permission required by or under any enactment;
  - the occupier must submit any plans or specifications for the alteration to the landlord for approval (provided that the condition binds the landlord not to withhold approval unreasonably) and that the work is carried out in accordance with such plans and specifications;
  - the landlord must be permitted a reasonable opportunity to inspect the work when completed; and
  - the occupier must repay to the landlord the costs reasonably incurred in connection with the giving of his consent.

- In cases where the landlord could reasonably withhold consent, he may instead give consent but impose a condition that the premises must be reinstated at the end of the lease.[561]

Similar (but not identical) provisions apply in relation to the duty on service providers.[562]

8

Costs and premiums

This chapter explains:

- when a tenant will be responsible for costs;
- the degree to which a landlord may demand costs;
- the meaning of 'reasonable costs';
- the restriction on recovering costs from residential tenants; and
- the difference between costs and premiums.

It does not deal with the rules on costs in litigation, which are different.

Responsibility for costs

The starting point: the tenant is not liable for costs

8.1 In the absence of some agreement or undertaking to meet the landlord's costs, the tenant is not liable.

Example

A tenant made three successive applications for consent to assign. For various reasons, none of the applications ultimately resulted in consent being granted. The landlord then sought payment of its costs. In two of the applications, the tenant's solicitors had given undertakings. However, they had not done so in the third, and the lease contained no express obligation to pay.

The court decided that there were no grounds for implying an obligation to pay costs into the lease. The landlord also argued that, by making an application, the tenant had impliedly agreed to pay. This was also rejected, so the tenant was not liable.[563]

It would probably have been reasonable, however, for the landlord to have asked for an undertaking to pay its reasonable costs. This is explained at 8.6 to 8.12.

Liability under the lease

8.2 The first step is to check the lease (and any other agreement with the tenant). If this contains an express obligation to pay the costs of an application, the tenant will be bound.

> ### Example
>
> A tenant might covenant 'to pay the landlord's reasonable costs of, and occasioned by, any application for consent under the terms of this lease, whether or not consent is granted, or whether granted subject to any lawful conditions'.

8.3 Any express obligation will be subject to the usual legal principles governing interpretation. Some initial guidance is given at 1.65 to 1.71.

8.4 An express obligation to pay costs may make it unreasonable for the landlord to demand a solicitor's undertaking in addition before dealing with an application for consent. This is explained further at 8.9.

8.5 If there is no express obligation, the landlord can consider asking for a specific promise to pay.

Undertakings for costs

8.6 A landlord, faced with an application from a tenant with no existing obligation to pay costs, will probably want to protect his position. As stated above, in the absence of an express agreement or undertaking, the tenant will not be liable.

8.7 The tenant may, of course, simply agree to meet the landlord's costs. However, the usual course is for the landlord to seek an undertaking from the tenant's solicitor. The extent to which the landlord can reasonably demand an undertaking is limited in two ways.

Limits on undertakings 1: reasonable costs

8.8   First, it is probably acceptable for a landlord to ask for an undertaking for *reasonable* costs, but to ask that it covers *all* costs is unreasonable.

Example

A tenant applied for consent to sublet part of an office block. The landlord asked the tenant's solicitors for an undertaking 'that [the tenant] will be fully responsible for all the costs, agents, solicitors and other associated costs' [sic]. The landlord would not deal with the application until the undertaking was given, and later estimated that the costs would be 'in the region of £4,500 plus VAT'.

The court decided that if a landlord could demand an undertaking at all, it could only cover *reasonable* costs. It was unreasonable to have asked for payment of *all* costs, and the sum of £4,500 plus VAT was an unreasonable amount.[564]

The court did not specifically sanction requests for undertakings for reasonable costs – it simply decided that asking for *all* costs was excessive. However, we consider that the landlord can ask for reasonable costs.

Although this case concerned a proposed subletting, the principle is likely to apply to other types of application.

Limits on undertakings 2: tenant already liable

8.9   Secondly, if the tenant is already liable it may not be reasonable to ask for an undertaking at all, at least if the landlord already has sufficient security for the costs and perhaps if the tenant is a strong covenant.

Example

A tenant applied for consent to assign, but the landlord's solicitor had delayed pending receipt of an undertaking for costs.

The lease contained an obligation on the tenant to pay the landlord's costs, and the landlord held a rent deposit which was more than enough to cover the potential liability. The court indicated that the request for an undertaking was unreasonable.[565]

Undertakings for costs 'whether or not the transaction is completed'

8.10 It is common for landlords to ask that undertakings will apply 'whether or not the transaction is completed' (or words to that effect). However, the courts have held that those words are not essential for the tenant to be liable if the application does not proceed.

Example

A tenant applied for consent. The tenant's solicitors had undertaken to 'pay [the landlord's] reasonable costs in relation to the licence'. Consent was then refused.

The tenant's solicitors contended that the undertaking did not apply, since no licence was ever granted and they had not undertaken to pay costs 'whether or not the transaction was completed'.

The court decided that the undertaking was wide enough to cover a situation where no licence was ever granted, and the absence of the extra words was not conclusive.[566]

8.11 If tenants or their representatives wish to limit their liability, that should be made clear in the wording of the undertaking itself. A tenant might, for example, undertake to pay the landlord's reasonable costs 'in connection with the *grant* of a licence'. That ought to give a more limited scope for liability than, say, an undertaking for costs 'in connection with the *application* for a licence'. The safest course would be to spell out that 'the tenant will not be liable for any costs in the event that no licence is granted', but this is unlikely to be acceptable to a landlord.

8.12 If an undertaking is given, the extent of the liability it imposes will be governed by legal principles. A detailed explanation of those principles is outside the scope of this book.

Reasonable costs

8.13 The types of expenses a landlord may incur, and the reasonable amount for the tenant to pay, is often a contentious issue. There are very few reported court cases dealing with this subject, so there is little definitive guidance.

8.14 However, the following points may be of assistance.

- The categories of expenses claimed by the landlord ought to be referable to the general principles on reasonableness. These are set out in chapter 7. In our opinion this means that costs may only be incurred on activities that relate to potentially reasonable grounds for refusal.

Examples

(1) It would not be reasonable for a landlord to seek costs incurred trying to persuade the tenant to surrender the lease. This is because a desire to obtain a surrender is not a good ground for refusing consent.

(2) Costs legitimately incurred assessing an incoming tenant's covenant strength or the structural viability of proposed alterations would be reasonable.

- The types of advisers from whom a landlord might (in appropriate circumstances) incur costs include:
  - *a solicitor*: for advising on the meaning of the lease and the landlord's duties, dealing with any necessary correspondence with the tenant (and/or his advisers) and drawing up a formal licence;
  - *a surveyor*: for advising on the impact of the application on the landlord's property interests, including, for example, the suitability of a proposed assignee or subtenant, the viability of proposed alterations or the effect of a change of use;
  - *a valuer*: if there is a contentious issue as to whether granting consent would have an impact on the capital value of the landlord's interest;
  - *counsel*: if there is a difficult point of law in issue; and

       – *an accountant*: for assessing the covenant strength of a proposed assignee or subtenant.

    Whether any of these costs would, in any given case, be deemed to be reasonable depends on the background circumstances and the details of the application.

- Whether any particular sum is, or is not, reasonable, will be a question of fact in every case, depending on all the surrounding circumstances. The more simple the application, and the less risk it poses to the landlord, the lower the reasonable costs are likely to be.

Examples

(1) A blue chip tenant of a long lease at a peppercorn rent seeks consent to sublet. The sublease will be for a short term, will be contracted out of the security of tenure provisions of the LTA 1954, and will contain a covenant to comply with the head-lease. This presents little risk to the landlord and no complicated legal or surveying issues. As such, the reasonable costs are likely to be low.

(2) A tenant proposes to sublet. The sub-rent is required to be at the passing rent level. The property is over-rented so the tenant will pay a reverse premium to the subtenant, who will gain protection under the LTA 1954. There are disputes about whether the tenant is entitled to give the reverse premium, the subtenant's financial strength and the impact on the capital value of the landlord's interest. The reasonable costs in this case are likely to be comparatively high.

    For an example of a sum which the court deemed unreasonably high, see 8.8.

- If consent is unreasonably withheld, the landlord will not be entitled to recover costs.[567]

8.15 This general guidance will only assist in cases where costs are required to be 'reasonable' (or words to that effect). If the tenant has an absolute obligation to pay all costs (perhaps under the terms of the lease), or no obligation at all, the position will of course be different.

Costs and residential property

8.16   A landlord's entitlement to costs is further restricted in the case of residential property. A tenant of a dwelling is only liable to pay a 'variable administration charge' under a lease to the extent that the amount is reasonable.

8.17   An administration charge includes an amount payable 'for or in connection with the grant of approvals under [the] lease, or applications for such approvals'. A variable administration charge is one that is neither specified in the lease nor calculated in accordance with a formula specified in the lease. A landlord's costs for dealing with an application under the lease will therefore fall within this definition, provided they are variable.

8.18   The landlord's demand for administration charges must contain a summary of the tenant's rights and obligations, failing which the tenant may withhold payment.[568]

Costs distinguished from premiums

8.19   The question of costs should not be confused with a landlord's demand for a premium. A demand for a premium is specifically outlawed in certain circumstances, which differ depending on the type of application. These circumstances are explained at 2.79 to 2.91, 3.35 to 3.51 and 4.50 to 4.64.

The meaning of 'premium' is explained below.

8.20   The statutes that prohibit premiums use the phrase 'fine or sum of money in the nature of a fine'. A fine is defined as including 'a premium or foregift and any payment, consideration, or benefit in the nature of a fine, premium or foregift'.[569] It has been described as something that goes 'irrevocably into the pocket of the landlord'.[570]

8.21   This is best illustrated by examples. The following demands have all been held to be a fine (or 'in the nature of a fine'):

- that a licensed 'free house' became tied to the landlord's brewery;[571]

- that the tenant paid the landlord his share of compensation for a compulsory purchase;[572]
- that the whole rent is paid in advance;[573]
- that the rent is increased;[574]
- that a service charge is increased;[575] and
- that a break clause is inserted (which when operated would give the landlord a financial benefit).[576]

8.22 On the other hand, a demand that the tenant gave the landlord a security deposit for the performance of a contract was held not to be a fine.[577]

Checklist: Costs

Use this checklist to identify whether the tenant will be liable for the landlord's costs, and to work through some of the other common issues.

- Does the lease contain an obligation to pay costs? If not, in the absence of an undertaking or other obligation to pay, the tenant will not be liable.
- If there is a costs clause in the lease, does it cover applications for consent? Are all costs covered, or only those that are reasonable?
- Has the landlord asked for an undertaking? When asking, landlords should remember only to seek *reasonable* costs, and that it may be unreasonable to ask for an undertaking at all if the tenant is already liable under the lease. See 8.8 to 8.9.
- What categories of costs is the landlord seeking to recover, and in what amount? See 8.13 to 8.15.
- In residential property, are the costs reasonable in amount? Is the demand accompanied by a statement of the tenant's rights and obligations? See 8.16 to 8.18.
- Does the request genuinely relate to costs, or is it a demand for a premium? See 8.19 to 8.22.

9

Granting consent

This chapter explains:

- what degree of formality is required to constitute consent;
- how consent might be granted inadvertently;
- whether consent can be revoked; and
- the key terms to consider in a formal licence.

What formalities are required?

9.1   As a starting point, it is important to know in advance what amounts to a valid consent. This will help landlords to avoid giving consent inadvertently. Equally, it will help tenants to know whether the landlord's behaviour or correspondence constitutes consent.

9.2   The answer to this question will often be very important: once consent is given it generally cannot be taken back, so the tenant will be free to go ahead (see 9.33 to 9.36). In the case of alienation, it will also have important consequences for any contract between the tenant and the proposed assignee or subtenant.

9.3   The first task is to check the terms of the lease, to see if there are any specific statements as to what constitutes consent. This will dictate the degree of formality required. The next task is to consider whether there are any other reasons why formal consent is needed.

Checking the lease

9.4   Generally speaking, the older the lease the less emphasis there is likely to be on formalities. Modern leases tend to include more detailed requirements, and often have a specific definition of the phrase 'Landlord's Consent' or similar. It is a matter in each case of reading and assessing the terms of the lease. Consider, for example, whether the lease indicates that a licence must be in writing or by deed.

9.5 Subject to that caveat, there are four broad categories into which most restrictions will fall. These are as follows.

No formal requirements

9.6 The lease may not say anything at all about how consents are to be expressed. An example of a clause of this kind is:

> 'Not to [...] *without the prior consent of the landlord'*.

This clause only requires that the landlord's consent is obtained before the restricted act (assignment, alteration, etc.) takes place. There is no need for the consent to be given in writing, much less for any particular formalities.

9.7 However, even if consent is given verbally it should be properly documented. This is not a legal requirement, simply good practice. A formal licence will provide a record for the future and can be used to define the limits of the consent and any other obligations the parties agree to take on.

Consent to be in writing

9.8 Most qualified or fully qualified covenants will fall into this category. The landlord's consent must be given in written form, but there are no other necessary formalities. An example of this kind of clause is:

> 'Not to [...] *without the consent of the landlord had and obtained in writing'*.

With a clause of this kind a purely verbal consent will not suffice.[578] There is an exception to this general rule, however, if the landlord gives consent orally as a means of fraud. Alternatively, the landlord might waive the need for writing either accidentally or deliberately. See 9.26 to 9.30.

The lack of any strict requirement for formalities means that consent can be given in correspondence,[579] including a letter from the landlord's solicitors.

Consent to be by deed

9.9 Some leases (particularly modern commercial leases prepared for institutional landlords) contain a definition of 'consent'. The definition will sometimes state that no consent is given until a deed or formal licence is completed. Any such formalities must be observed.

9.10 However, the need for a deed is not the same as a separate requirement that (in an alienation case) an assignee should give a direct covenant to observe the terms of the lease. That alone does not mean that consent cannot be given in the form of a letter,[580] but the assignee will still need to give the direct covenant or there will be a breach of the lease.

Absolute covenant

9.11 Absolute covenants do not fall into any of the three categories set out above. This is because they create an absolute prohibition against the tenant assigning, altering the premises, etc. Therefore, such clauses will not provide any mechanism for the landlord to give consent.

9.12 If the parties agree that the tenant should have permission to go ahead in spite of the prohibition, a formal licence should be drawn up.

Other reasons for formal consent

9.13 The next task, once the lease has been checked, is to consider whether there is any other reason why a formal licence might be required.

9.14 Formal consent will often be needed under a contract of sale between the tenant and a prospective assignee or subtenant. Such contracts are often made conditional on the tenant obtaining the landlord's consent, with completion of the sale only being triggered when consent is granted. In these circumstances, it might be a condition of the contract that consent is documented, and in any case it would be good practice to do so.

Inadvertent consent

9.15 Assessing the lease terms will tell you what degree of formality is ordinarily needed to constitute consent. However, it is possible that the landlord might give consent in the eyes of the law without necessarily meeting those formalities (or without intending to do so). It is therefore important to keep a close watch on exactly what takes place between the landlord, tenant and any third party.

9.16 The courts will look at the objective meaning of the landlord's actions, not the subjective intention behind them. Therefore, there is ample scope for a landlord to give consent without necessarily meaning to do so.

Giving consent in writing

9.17 Where the lease dictates no formalities or only requires writing (as to which see 9.6 to 9.8), the pitfalls for landlords come in two general categories: giving consent in correspondence; and relying on the 'subject to licence' maxim.

Consent given in correspondence

9.18 In deciding whether consent has been given, the courts will look at the correspondence as a whole.[581] Consent may be given where the landlord simply indicates a willingness to grant consent, and even though he intends that a formal licence should follow.

Examples

(1) A landlord's solicitors wrote to the tenant saying that the landlord was 'quite willing to consent' to an underlease and that they would forward a licence provided the tenant would meet the costs. The court held that the correspondence was clear enough to constitute consent, and the licence was meant to be a mere formality.[582]

(2) A tenant applied for licence for a change from shop use to a restaurant. The landlord's agents wrote to the tenant saying that 'subject to the solicitors drawing up the said deed of variation, your landlord would have no objection to a change of use'. There was no need for a deed, only for written consent. The court held that consent had been given.[583]

Consent given 'subject to licence'

9.19 In the context of consents, this phrase does not carry the weight normally attributed to it.

9.20 The usual rule is that the words 'subject to contract' will prevent a binding arrangement coming about before the legal formalities are completed. However, the courts have decided on several occasions that this does not apply to landlords' consents. Where the lease only requires consent in writing, a letter that gives consent may suffice, even if marked 'subject to licence'.

Examples

(1) A tenant applied for consent to make alterations. The landlord's agent wrote to the tenant saying 'the freeholder … gives consent for the works'. The letter went on to outline three conditions: formal licence, payment of costs and the tenant obtaining statutory approvals. The letter was also marked 'subject to licence'. The court decided that consent had been given by that letter.[584]

(2) A tenant applied for consent to assign. The landlord's surveyor wrote to the tenant saying that she had received satisfactory references. That was the only outstanding request the landlord had made. The letter went on to state that solicitors had been instructed to issue a draft licence. Again, the letter was marked 'subject to licence', and again the court decided that it sufficed as consent.[585]

Avoiding giving consent in correspondence

9.21 In all of the cases where consent was held to have been given, the letters either stated expressly, or gave the clear impression, that consent was forthcoming. If a landlord avoids that trap, consent will not be granted. For example, if the landlord states that he 'is considering granting consent and will forward a draft licence', he is unlikely to be held to have granted consent.

9.22 Before the LTA 1988 came into force, it was held that where a landlord stated at the outset that 'no consent will be granted until the execution of a formal licence to assign', no consent had indeed been given.[586] This is likely to remain the case where the Act does not apply. Where the Act does apply (see 1.27 to 1.31), there is a risk that the landlord might be held not to have granted consent within a reasonable time. We suggest that better practice is simply to grant consent, with the intention of recording the consent in a licence later if appropriate.

9.23 If the lease requires that any consent is given by deed, it is much less likely that the landlord's correspondence will be deemed to constitute consent. However, it remains possible that the landlord's conduct overall might lead a court to declare that consent has been given.

Consent given 'without prejudice'

9.24 Where a landlord may not withhold consent unreasonably, this heading will be meaningless. The landlord is required either to give consent, or to withhold it on reasonable grounds.

9.25 As explained at 7.45, the landlord may give consent without prejudice to his right to forfeit the lease (if the landlord believes the tenant is in breach of covenant when the application is made). However, this must be explained carefully in the landlord's response. The general use of the heading 'without prejudice' is only relevant where negotiations to compromise a dispute are taking place.

Giving consent by conduct

9.26 Where the lease demands that any licence shall be in writing or by deed, there are circumstances in which the courts will, nevertheless, declare that consent has been given by the landlord's conduct alone. It is possible to put these circumstances into two categories: where the landlord is guilty of fraud; and where the landlord waives the requirement for consent in writing.

Landlord guilty of fraud

9.27 The courts will not allow landlords to rely on the legal requirement for a written consent if there is evidence of fraud.[587] This is consistent with the legal principle that 'fraud unravels everything'.

9.28 Therefore, if a tenant is able to prove fraud on the part of the landlord (or, in the words of one judgment, that oral licence was 'used as a snare')[588] the court will uphold the oral consent despite the terms of the lease.

Landlord waives requirement for consent in writing

9.29 This second category involves the landlord acting in such a way as to make it unjust to allow him to enforce the strict terms of the lease.

Example

A tenant wanted consent to assign a lease. A meeting was arranged between the landlord, the tenant and the proposed assignee. There was a discussion about the proposed use of the premises, some alterations and the references to be supplied. The parties reached agreement on all the issues, and the landlord said that he would accept the proposed assignee as his tenant. The assignee then entered into obligations with the tenant.

The judge held that the landlord had waived the requirement for written consent.[589]

9.30 In our opinion, the landlord may be prevented from arguing that he had not given consent if:

- he gives consent orally; and
- his conduct leads the tenant to go ahead on that basis; and
- the tenant relies on the landlord's conduct and accepts the oral consent.

However, it will depend on all the circumstances.

Inadvertent consent with absolute covenants

9.31 Where the lease contains a complete prohibition on a certain act, it is much less likely that a landlord will give permission inadvertently. However, it is not impossible. The tenant might, for example, argue that the landlord has released the restriction altogether.

9.32 A discussion of this area of law is outside the scope of this book.

Can consent be revoked?

The general rule

9.33 The general rule is that, once granted, consent cannot be withdrawn. This is the case even if the landlord later discovers some fact that would have justified withholding consent.

Example

A tenant wanted to assign a lease of a bakery. The landlords gave consent, but then discovered that the proposed assignee had convictions for food hygiene offences. They said that the consent was therefore withdrawn. The court decided that the tenant was not at fault: he had not known about the conviction, and in any case it was for the landlords to make such enquiries as they thought fit. The tenant had acted on the consent, by contracting with the proposed assignee, so it would be unfair to allow the landlords to withdraw it.[590]

Exception in the case of fraud and misrepresentation

9.34 There is an exception where the tenant has obtained the licence by fraud. The landlord may ask the court to 'set aside' the licence. This does not mean that an assignment that took place on the basis of the consent is ineffective; it is simply treated as though it was done without consent.[591] This means that the landlord may use the remedies explained in chapter 10.

9.35 It is likely that a misrepresentation in the application for consent would lead to a similar result. Any licence given by the landlord extends only to the permission actually given and nothing else, unless the licence states otherwise.[592] Therefore, a tenant who misrepresents the nature of the application does not receive consent for the true proposal.[593] However, if the misrepresentation is about an immaterial matter, it will be irrelevant.[594]

Exception where there is express provision in the lease or the grant of consent

9.36 Leases occasionally contain clauses that allow a landlord to revoke a consent between the time when it is granted and the time it is put into use (for example, when a disposition is completed or alterations begin). This is likely to apply if the landlord discovers something which justifies the decision to revoke. We consider that in these cases a landlord can withdraw his consent.

In the case of consents for continuing acts, such as consent to keep a sign in place or perhaps consent to change use, it may even be possible to revoke a consent after the tenant has acted on it if the landlord expressly reserves the right to revoke it when granting consent.[595] However, if the landlord is required to act reasonably, it is doubtful that it would be reasonable to reserve that right.

The contents of the licence

9.37 Once the parties have agreed that a formal licence should be issued, they and their advisers will need to decide on the terms it will contain. Some of the issues the parties might wish to consider are listed below, categorised according to the type of application the tenant has made.

9.38 This is not intended to be a guide to drafting or negotiating the legal content of the licence, nor do we make any detailed comments on the issues covered. The issues relate to all covenants, whether absolute, qualified or fully qualified; we do not consider here whether it would be reasonable (in the context of a fully qualified covenant) to insist on any of them as a term of consent. Questions of reasonableness are discussed in chapter 7.

9.39 Landlords in particular should be aware that requiring a formal licence as a condition of consent may be unreasonable, or the requirement may simply be ineffective. See 7.46 to 7.50.

Assignment or subletting

Direct covenant

9.40   An assignee or subtenant will often give the landlord a direct covenant to observe the terms of the lease. In the case of a subtenant, the direct covenant usually excludes the obligation to pay the head-rent but includes an obligation to comply with the terms of the sublease.

Guarantee

9.41   If the lease is a 'new' tenancy, the tenant may be required to enter into an Authorised Guarantee Agreement (AGA). This is likely to be incorporated into the licence. AGAs are explained at 7.133 to 7.137.

Most tenancies granted in or after 1996 are 'new' tenancies. See Appendix I for more details.

9.42   The assignee may also provide a guarantor, whose obligations may be recorded in the licence.

Rent deposit

9.43   Is the incoming tenant to provide a rent deposit? If so, on what terms is it to be held?

Alterations

Recording the works

9.44   If the licence is to serve as a record of what has been agreed, it is important that the works are properly described. Thought should be given to whether plans alone, annexed to the licence, will suffice. A written description of the works, even a very brief summary, is very helpful for future reference.

The conduct of the works

9.45   Tenants will usually give a covenant to obtain any necessary consents (for example planning permission), and to carry out the alterations 'in a good and workmanlike manner' (or similar) and in accordance with all applicable regulations. Does the landlord want the right to inspect the works as they progress? If so, how frequently and who is to be responsible for the costs?

Whether the tenant must do the works

9.46 Will the tenant be *obliged* to carry out the works, or does he merely have *permission* to do them? This may affect future rent reviews, if the review clause provides that works undertaken as part of an obligation to the landlord are to be taken into account, but other works are to be disregarded. If the tenant is to be *obliged* to do the works, he may therefore end up paying for them twice.

Whether the alterations are to be reinstated

9.47 Tenants often agree to reinstate alterations at the end of the lease. If the landlord wishes for this to happen, it must be expressly agreed. In the absence of an express obligation, the tenant is not required to reinstate (as the alterations are deemed to have become part of the premises).

The terms of any reinstatement clause

9.48 The tenant might agree to reinstate the alterations:

- automatically (i.e. without prior notice) before the end of the lease;
- only if the landlord gives notice requiring reinstatement; or
- automatically, unless the landlord gives notice not to do so.

9.49 If the parties intend to use the second alternative, thought should be given to the form and timing of the notice.

- If nothing is agreed about the form of notice there is scope for dispute about what is required. No particular form will be needed, but the landlord must make it clear that he requires the premises to be reinstated. In one case, it was held that a requirement for reinstatement in a schedule of dilapidations sufficed.[596]
- As to timing, it is thought that notice must be served before the lease ends.[597] However, there appears to

be no rule that it must be served early enough for the tenant to complete the works before the lease ends.[500]

9.50 Similar issues arise with the third alternative. For example, what is the position if the tenant has already done the work before the landlord gives notice?

9.51 Given the degree of doubt, the parties may find it useful to make specific provision for these issues in the licence.

Security for reinstatement

9.52 Tenants sometimes agree to give the landlord some security for their obligation to reinstate. The parties may wish to consider whether security should be provided and, if so, how much and in what form.

Schedule of condition

9.53 A common cause of dispute, where alterations are to be reinstated, is the exact state of the property beforehand. Completing a schedule of condition before the works begin, and attaching it to the licence, would help to avoid this problem. More realistically, 'before' and 'after' drawings should be attached.

Compensation for improvements

9.54 In certain circumstances, a tenant has the right to compensation at the end of the lease for any improvements he has carried out. This arises under the LTA 1927, and is discussed at 3.88 to 3.92.

9.55 The process begins with the tenant giving notice of intention to make improvements. Because the notice may not need to refer expressly to the LTA 1927,[599] it is possible that a simple application for consent might qualify. Therefore, if the works are not qualifying improvements and it is not intended that the tenant should be compensated, the landlord may wish to record that fact in the licence.

Effect on rent

9.56 The parties should consider how the alterations will affect the value of the property on future rent reviews. The licence could be used to document an alternative arrangement.

Change of use

Whether the new use is personal to the current tenant

9.57   The parties should consider whether the new use will apply for the rest of the term of the lease (regardless of any assignments or sublettings), or simply be personal to the current tenant. The agreement should be recorded clearly in the licence to avoid future disputes.

9.58   If nothing is expressly agreed, the question is likely to depend on all the background circumstances. However, if the tenant intends to spend money converting the property to the new use and establishing the new business, it is more likely that consent will be construed as lasting for the whole term.[600]

9.59   Even if the consent is expressly agreed to be personal to the current tenant, it may still bind a buyer of the landlord's property unless the licence specifically says otherwise.[601]

Planning consent

9.60   The landlord is likely to want the tenant to confirm that any necessary planning consent has been (or will be) obtained.

Positive or negative covenant

9.61   Will the tenant be *obliged* to use the premises for the new purpose, or simply *permitted* to do so? In other words, will the tenant have the option of not using the premises at all? See 4.9 to 4.12 for more information.

Revocation of old use

9.62   Will the tenant have the option of reverting to the old use, or may he only use the premises for the new purpose?

Exclusion of warranty

9.63   It is common for landlords to incorporate a provision excluding any warranty that the premises can lawfully be used for their new purpose. In fact, this is unnecessary because no such warranty is implied.[602]

Effect on rent

9.61 The tenant's freedom to use the property may have a significant impact on the rent at review. This will depend on the terms of the review clause. Therefore, both parties ought to consider how the change will affect future valuations. If necessary, the licence can be used to record an alternative agreement.

For example, whether the change is personal or binds future tenants is likely to be important. If the rent review clause in the lease provides that a particular use is to be assumed, that will reduce the impact on review.

Checklist: Granting consent

Use this checklist to work through the key points on the granting of consent.

- Does the lease dictate formalities for a licence? See 9.4 to 9.12.
- Are there other reasons why a licence is needed? See 9.13 to 9.14.
- Has consent already been given inadvertently? See 9.15 to 9.32.
- If consent has been given, can the landlord revoke it? See 9.33 to 9.36.
- What other terms might be important? See 9.37 to 9.64.

10

Enforcing rights

This chapter is in three parts:

- The first part covers remedies for landlords against tenants and explains:
  - how to check whether a tenant is in breach of covenant;
  - when and how the landlord can take action against the tenant by claiming damages, seeking an injunction or forfeiting the lease; and
  - how to avoid losing the right to forfeit.
- The second part covers remedies for tenants against landlords and explains:
  - what the tenant can do if he thinks consent has been wrongly withheld; and
  - when the tenant can claim damages for a wrongful withholding of consent.
- The third part explains when surveyors might be liable to landlords or tenants.

Remedies for landlords

Preliminary: checking for a breach of covenant

10.1 Assessing whether a tenant is in breach of a covenant against alienation, alteration or change of use has two aspects:

- Has the tenant done something which is restricted by a covenant in the lease? Chapters 1 to 4 give some guidance on interpreting covenants, but it will often be wise to seek legal advice.

- If he has, is the landlord able to take action?
  - *Absolute covenants:* The landlord will usually be able to take action if there is a complete prohibition, unless he has given consent. If the landlord has consented to the act, he will not be able to seek a remedy.
  - *Qualified covenants:* The landlord will also usually be able to take action if the covenant is a qualified one that is not made fully qualified (i.e. covenants against changing use and limited other covenants explained in Appendix 2), unless he has granted consent or demanded an unlawful premium.

    As explained at 10.55, if the landlord has demanded an unlawful premium, the need for consent falls away. The tenant is entitled to go ahead: he will not be in breach of the lease or need consent.
  - *Fully qualified covenants:* If the covenant is fully qualified, then whether the tenant is in breach of covenant will depend on the landlord's conduct.

    As explained at 10.51 to 10.54, if the tenant applies for consent, and consent is unreasonably withheld or granted subject to an unreasonable condition, the need for consent falls away. The tenant is entitled to go ahead without breaching the lease or needing consent.

    Again, the landlord may have inadvertently granted consent.
  - *Loss of rights:* In exceptional cases, the landlord may be unable to enforce any type of covenant (whether absolute, qualified or fully qualified) because he has either waived the breach or impliedly released the covenant itself. This is usually because he has accepted a breach for a long time, but is comparatively rare.

For an explanation of absolute, qualified and fully qualified covenants, see 1.6 to 1.15.

For details of when qualified covenants are made fully qualified and when a premium can lawfully be demanded, see 2.79 to 2.91, 3.35 to 3.51 and 4.50 to 4.64.

For an explanation of when a landlord may have inadvertently given consent, see 9.15 to 9.32.

10.2 If the tenant is in breach of covenant, the landlord has a number of possible remedies. It is beyond the scope of this book to deal with these remedies fully, but this section gives some basic guidance to help landlords choose their approach, and flags up some important points.

Damages

10.3 If the tenant is in breach of covenant, the general rule is that the landlord is entitled to damages. 'Damages' means compensation to put him in the position in which he would have been if the covenant had been observed. He is entitled to damages in addition to an injunction or forfeiture of the lease. However, if he is awarded some other remedy this is likely to limit any loss he has suffered, so the damages will be reduced.

10.4 In some cases the landlord can demand a premium in return for giving consent. These are:

- an absolute covenant;
- a qualified covenant against change of use which involves structural alterations;
- limited other qualified covenants which the law does not make fully qualified and where the law does not prohibit a premium; and
- some fully qualified covenants where the lease specifically allows premiums.

See Appendix 2 for details of when covenants are made fully qualified and when premiums are prohibited.

See 2.88 to 2.90, 3.45 to 3.50 and 4.62 to 4.63 on when the lease specifically allows premiums.

10.5 In these cases, if the tenant breaches the covenant the landlord will have been deprived of the opportunity to demand a premium. His damages can include a sum to compensate him for the loss of that opportunity.[603] The damages will be the sum that a reasonable landlord and a reasonable tenant would have agreed as the price of the consent.

10.6 If the tenant should have applied for consent but failed to do so, the tenant will be in breach of covenant, even if consent could not reasonably have been refused if it had been asked for.[604] However, any damages are likely to be nominal.

Superior landlord

10.7 If a subtenant covenanted not to do something without a superior landlord's consent, the superior landlord will be able to claim damages from the subtenant if he goes ahead without it. This can be as well as or instead of the immediate landlord. This applies whether or not the superior landlord was a party to the sublease; it is only necessary that the covenant was clearly given for the superior landlord's benefit.[605]

> *Example*
>
> A sublease contains a covenant '*not to assign, underlet or part with possession of the premises without the consent of the superior landlord*'. The superior landlord was not a party to the sublease. The subtenant assigns without obtaining the consent of the superior landlord.
>
> The covenant was clearly inserted for the benefit of the superior landlord, and so he will be able to claim damages.

Injunction

10.8 Injunctions can be divided into two types:

- positive, or 'mandatory', injunctions; and
- negative, or 'prohibitory', injunctions.

10.9 Positive injunctions order a tenant to do something. For example, a tenant might be ordered to reinstate unauthorised alterations. This type of injunction can often be hard to get, particularly if the landlord has delayed. It is therefore important to get legal advice early on.

10.10 Negative injunctions stop a tenant from doing something. For example, a tenant might be ordered not to use premises as a restaurant. This type of injunction is much easier to obtain. Even so, if the landlord delays a lot or leads the tenant to think that the breach is accepted, or if an injunction would be oppressive, it may well be refused.

10.11 If the landlord discovers that the tenant is about to breach a covenant, he can apply for an 'interim' injunction to prevent the tenant from doing so. However, he will usually have to undertake to compensate the tenant for any loss the tenant suffers if it turns out that the injunction should not have been granted. This can make it a risky strategy for the landlord if it is not a clear-cut case. Nonetheless, there are risks in not applying: sometimes the court may be less willing to grant a final injunction at the end of the case.

Example

If a landlord refuses consent to some alterations, and then sees scaffolding going up at the premises, he can apply for an interim injunction to order the tenant to stop the work. This injunction will usually last until the court has made a final decision on whether the tenant should be prevented from carrying out the alterations.

If the court finally decides that the landlord's refusal of consent was reasonable and the tenant has no right to make the alterations, it is likely to order the tenant to do any reinstatement work that is required.

If the landlord had not applied to halt the works at an early stage, the court might have been more reluctant to order the tenant to reinstate.

However, if the court finally decides that the landlord's refusal was unreasonable, the landlord will have to compensate the tenant for any loss suffered as a result of the delay to his works.

10.12 If the tenant should have applied for consent but failed to do so, the tenant will be in breach of covenant. This is true even if consent could not reasonably have been refused, although in that case the landlord will be less likely to get an injunction.

10.13 Particular points to note about injunctions are:

- *Keep open clauses*: The court will almost never make an injunction requiring a tenant to carry on a business when he does not want to.[606] Instead the landlord will usually be awarded damages. This can create a difficult decision for landlords.

Example

A lease contains a covenant not to use the premises other than as a supermarket without landlord's consent. It also contains a covenant to keep the premises open for trade during normal business hours. In breach of covenant, the tenant ceases trading as a supermarket and assigns the premises to an entertainment retail store instead. This is in direct competition with another of the landlord's tenants next door but one, who sells music CDs only.

The landlord would like to stop the competition, but it is also important to him that the premises are not left empty. If he applies promptly for injunctions, he will probably be granted an injunction preventing the new tenant from using the premises as an entertainment retail store. But he will not be granted an injunction requiring the new tenant to keep the premises open as a supermarket. So he will have to decide whether to accept the unauthorised use or to have the premises left empty unless a new supermarket tenant can be found.

- *Assignments and sublettings*: If the tenant has assigned or sublet in breach of covenant, it is possible to get an injunction requiring the lease to be assigned back or the sublease to be surrendered.[607] However, the landlord is only likely to be able to get an injunction if it can show that the assignee or subtenant knowingly and intentionally induced the breach of covenant by the tenant:

  'It is not sufficient merely to show that [the subtenant] was muddle-headed or illogical in what it thought and did or that it had the means of knowledge that a breach of covenant was being committed (leaving aside questions of recklessness). Actual knowledge of the breach, and the intention that the breach be committed, must be established.'[608]

Even then, there may be other reasons to refuse an injunction.

- *Application by the tenant to discharge or modify the covenant*: Tenants of long leases may apply to the Lands Tribunal for a covenant restricting use or alterations to be discharged or modified. This is explained at 4.68 to 4.73. If the landlord applies for an injunction, the tenant can apply for an order staying the landlord's proceedings while he makes an application to the Lands Tribunal.[609]

Superior landlords

10.14 If a subtenant has covenanted to obtain a superior landlord's consent before doing something and fails to do so, the superior landlord has the same rights to seek an injunction as the intermediate landlord does. This applies whether or not the superior landlord was a party to the sublease; it is only necessary that the covenant was clearly given for the superior landlord's benefit.[610] See the example at 10.7.

10.15 In the case of restrictions on the use of the premises, the superior landlord will often be able to obtain an injunction to prevent the subtenant from breaching the restriction even if the covenant was given only by the tenant and not by the subtenant.

Forfeiture

10.16 Almost all leases contain a clause allowing the landlord to terminate (or 'forfeit') the tenancy prematurely if the tenant breaches a covenant. Forfeiture can therefore be a useful remedy if the tenant's breach has done such harm that the tenant needs to be removed. It can also be attractive if the landlord wants the property back. However, usually the tenant is permitted to stay, and forfeiture simply enables the landlord to enforce covenants indirectly: see 10.30 to 10.32.

10.17 Legal advice should be taken before forfeiting a lease, but a few key points are set out here. These apply to non-residential tenancies; for residential tenancies, see 10.35 to 10.46. Different rules again apply to agricultural tenancies, but those are beyond the scope of this book.

Warning notice

10.18 If the landlord wants to forfeit the lease for a breach of a covenant against alienation, alterations or change of use, he must first serve a warning notice. This is known as a s. 146 notice because the obligation to serve it arises under s. 146 of the LPA 1925. The notice tells the tenant what he has done wrong, and gives him a chance to put matters right, or at least to consider his position.

10.19 The notice should be addressed to the tenant. This means that if there has been an assignment, the notice must be addressed to the assignee. This is the case even if the assignment was in breach of covenant,[611] because an assignment in breach of covenant still transfers the lease to the assignee.

10.20 Legal advice should be taken to make sure the notice is valid and sent to the right person.

Waiving the right to forfeit

10.21 If the landlord or his agent:

- knows of the facts which allow him to forfeit the lease (which will usually be the facts giving rise to the breach of covenant); and

- does an unequivocal act which recognises the continued existence of the lease (explained at 10.25 to 10.26),

then he will not be able to forfeit afterwards, unless he has new grounds to do so. He will be regarded as having made a decision not to forfeit the lease and to use his other remedies instead, and he cannot go back on that decision.[612]

10.22 This is the case even if:

- the landlord did not intend to lose the right to forfeit;
- the lease contains a provision saying that nothing the landlord does will waive the right to forfeit;[613]
- the landlord did not appreciate the legal effect of the facts he knew;[614] or
- his agent knew relevant facts and did not pass them on to him.[615]

The landlord and his agents (including his surveyor) therefore need to be very careful as soon as they find out facts which may give rise to a breach of covenant. If the landlord finds out that the tenant has assigned, sublet or made alterations in breach of covenant, and subsequently does something which unequivocally recognises that the lease is still continuing, he will not be able to forfeit.

10.23 The position is different where the breach of covenant relates to a restriction on use. Unauthorised use, such as breach of a covenant prohibiting residence on the demised premises, is usually a 'continuing breach'.[616] This means that a fresh breach arises every day until the unauthorised use ceases, and the landlord can rely on a later breach even if he waives the right to forfeit for an earlier one.

However, if the unauthorised use is the consequence of a 'once and for all' breach, the continuing breach will not assist the landlord. For example, if the tenant sublets part, in breach of a covenant not to do so (a once and for all breach) and of a covenant to use only as a single private dwelling-house (normally a continuing breach) acceptance of rent with knowledge of the subletting will waive the right to forfeit for both breaches.[617]

10.24 Under this principle, the landlord only waives his right to forfeit the lease. However, it is also possible to waive the *breach* (which means that the landlord cannot use any of his remedies in respect of that breach), or to waive the *covenant* (which means that the landlord will not be able to take any action in respect of the current breach or any future breach).

Avoiding waiver

10.25 The most common act that will prevent the landlord from forfeiting is acceptance of, or even demand for, rent. This applies:

- whether the rent was accepted or demanded by the landlord or by his agent; and
- even if the rent is accepted or demanded 'without prejudice to the right to forfeit'.[618]

However, it only applies to rent that falls due after the landlord discovers the breach. So if the tenant assigned in breach of covenant on 24 March but the landlord did not find out until 2 June:

- the landlord can safely accept rent which fell due on 25 March (even if the tenant does not pay it until 3 June); but

- if he accepts rent which fell due on 24 June, he will not subsequently be able to forfeit.

It also only applies to rent which is paid before the forfeiture. If the landlord begins a claim for possession and serves it on the tenant on 7 September, he will then be able to accept the rent which fell due on 24 June.[619]

If rent is paid by standing order or direct debit, the landlord will probably still be able to forfeit if he returns the rent promptly after becoming aware of the payment, at least where rent has not been demanded and the tenant has been told that rent will not be accepted.[620]

If the landlord draws on a rent deposit for rent which fell due after he knew about the breach, he will probably lose the right to forfeit.

If the landlord accepts payment from a guarantor for rent which fell due after he knew about the breach, he will probably lose the right to forfeit.[621]

10.26 Other acts that will prevent the landlord from forfeiting include:

- an unequivocal grant of a licence to assign or sublet, although a landlord can deal with an application if he expressly reserves the right to forfeit;[622]

- registration of notices;

- an offer to purchase the tenant's interest;[623] and

- commencing and maintaining proceedings for an injunction. However, a landlord can seek an injunction and possession in the alternative, and defer making his decision as to whether to forfeit or not until trial.[624]

10.27 The landlord and his agents should therefore be careful not to do any of these things after they find out about the facts giving rise to the breach of covenant, until the lease has actually been forfeited. After the forfeiture has taken place, the landlord cannot change his mind, and so the right to forfeit cannot be waived.[625]

Effecting the forfeiture

10.28 The landlord can carry out the actual forfeiture either by starting a court claim for possession and serving it on the tenant or by 'peaceably re-entering'.

Peaceable re-entry means physically entering the premises, using reasonable force if necessary to gain entry. Two points to note are:

- It is important that the premises are empty at the time because if there is someone physically present who opposes the entry, the landlord will be committing a criminal offence if he uses force.[626] This includes violence to property as well as to people. Peaceable re-entry is therefore usually effected in the early hours of the morning – unless the tenant operates a nightclub or the like. It is also usually effected without warning, to make it less likely that the tenant will engage 24-hour security.

- It is also important that nobody is lawfully occupying any part of the premises as a residence, whether or not they are physically present at the time of the re-entry. Otherwise, the landlord will again be committing a criminal offence if he peaceably re-enters.[627] Even if the residential part has a separate entrance which will be left open, the landlord must still forfeit by bringing court proceedings.[628]

10.29 The effect of forfeiting is to end the lease, subject to the right to claim 'relief'.

Relief from forfeiture

10.30 The court has wide powers to grant 'relief' from forfeiture, which means either reinstating the lease or creating a new lease for a subtenant or a lender (if the tenant has mortgaged the lease). Relief will usually be granted if the landlord is put back into the position that he would have been in if the breach had not been committed (for example, any unauthorised alterations being reinstated). Therefore, forfeiture is more usually an indirect method of enforcing covenants.

10.31 In cases where the tenant should have applied for consent but failed to do so, it will be important to consider whether consent could have been reasonably refused in any event.

10.32 Some early cases held that this fact was irrelevant and the tenant was refused relief from forfeiture.[629] However, these cases pre-date the court's wide power to grant relief from forfeiture, and we consider that in many cases relief is now likely to be granted. There may not even be any requirement that the breach is remedied, at least if the failure to ask for consent was not deliberate and sometimes even in the case of deliberate breaches.[630]

Superior landlords

10.33 If a subtenant has covenanted to obtain a superior landlord's consent before doing something and fails to do so, the superior landlord has no direct right to forfeit the subtenancy.

10.34 However, if there is also a breach of the intermediate tenant's lease, the superior landlord may be able to forfeit that lease. Assuming that the subtenancy was granted before the intermediate lease, this will automatically terminate the subtenancy as well. The subtenant may be able to seek relief from forfeiture (explained immediately above) if the intermediate tenant does not, even if he is only a subtenant of part.

Long residential leases

A long lease is usually one for 21 years or more. The full definition is beyond the scope of this book.

10.35 Similar rules apply to long residential leases. However, before the landlord may serve a warning notice under s. 146 of the LPA 1925 he must either (a) obtain a determination from the Leasehold Valuation Tribunal or the court that a breach of covenant has occurred, or (b) obtain an admission of breach from the tenant.

10.36 If the tenant admits the breach, a s. 146 notice can be served immediately. If the landlord needs a determination, the notice cannot be served until 15 or more days after the determination has become final, i.e. when the time for bringing an appeal is over or any appeal is abandoned.[631]

10.37 As the premises are residential, forfeiture must take place by court proceedings (not peaceable re-entry), as explained at 10.28.

Assured and assured shorthold tenancies

> Assured and assured shorthold tenancies are types of private sector residential tenancy created after 15 January 1989. For more details, see Appendix 1.

10.38 A completely different regime applies to assured and assured shorthold tenancies. Again the landlord must serve a warning notice, but this notice is required by s. 8 of the HA 1988 and so takes a different form from a s. 146 notice. The landlord may bring possession proceedings after two weeks. Whether or not to grant possession is in the court's discretion.

Secure tenancies

> A secure tenancy is a type of public sector residential tenancy. For more details, see Appendix 1.

10.39 Different regimes apply to secure tenancies, depending on whether the tenancy is a periodic tenancy or one for a fixed term:

- If the tenancy is periodic, then the landlord must serve a warning notice. In this case, the notice is served under s. 83 of the HA 1985. The date on which the landlord can bring possession proceedings will vary according to the type of tenancy, but cannot be less than 28 days for breaches of covenants against alienation, alterations or change of use. Whether or not to grant possession is in the court's discretion.

- If the tenancy is for a fixed term, then the same rules apply as those for non-residential tenancies, so the landlord must also serve a warning notice under s. 146 of the LPA 1925. However, when the landlord applies to court, the court will not order possession; instead the court will make an order ending the fixed term tenancy. After that, a statutory periodic tenancy will arise. The same rules then apply as for any other periodic secure tenancy (explained in the previous paragraph).

The landlord may serve a warning notice under s. 83 before the fixed term is ended and start a claim for possession at the same time as seeking an order ending the fixed term tenancy.

- As explained at 2.62 to 2.71, if the tenant assigns, sublets or parts with possession of the property, sometimes the tenancy will cease to be secure. In these cases, the landlord can simply serve a notice to quit. The same will usually apply if the tenant moves out of the property. When the notice expires, the landlord can apply to court for an order for possession, and in this case the court must grant the order.

Rent Act tenancies (1): breach of covenant

A Rent Act tenancy is a type of private sector residential tenancy usually created before 15 January 1989. For more details, see Appendix 1.

10.40 The same rules apply to protected fixed term tenancies (i.e. tenancies where the original contract for a fixed term still applies) as those for non-residential tenancies, so the landlord must serve a warning notice under s. 146 of the LPA 1925. But because the tenancy is residential, the tenancy must be ended by a court order: see 10.28 and 10.42.

10.41 For protected periodic tenancies (i.e. tenancies where the original contract for a periodic tenancy still applies), the landlord can do the same, but it will usually be easier and safer simply to end the protected periodic tenancy by serving a notice to quit. However, termination of the protected tenancy does not entitle the landlord to possession. After the protected tenancy is terminated, a statutory tenancy arises if, and for so long as, the tenant continues to live at the property.

10.42 To end a statutory tenancy and get possession, the landlord needs to start a court claim, but he does not need to serve any warning notice in advance.

- If the preceding protected tenancy has already ended, for example if a fixed term has already expired, the landlord can start a possession claim immediately.

- If the preceding protected tenancy is to be terminated by forfeiture, the landlord can start a possession claim to end the statutory tenancy at the same time as the claim to forfeit.

- If the preceding protected tenancy was periodic and ended by notice to quit, the landlord will have to wait until the notice to quit has expired before starting a possession claim to end the statutory tenancy.

Rent Act tenancies (2): alienation

10.43 The landlord can get possession from a Rent Act tenant who has assigned or sublet even where there is nothing in the tenancy agreement restricting the tenant's rights to assign or sublet. This is in addition to the procedure described at 10.40 to 10.42 to take action for a breach of a covenant in the lease. The right is known as Case 6.[632]

10.44 Case 6 applies where the tenant:

- has assigned or sublet the whole of the dwelling-house; or has sublet part of the dwelling-house, the remainder already being sublet; and

- has done so without the landlord's consent.

Note that Case 6 does not apply if the tenant has only assigned or sublet part of the dwelling-house.

Note also that there are date restrictions, but these are unlikely to be of continuing relevance except for very historic assignments and sublettings.

10.45 If Case 6 applies, then the court may make an order for possession if it is reasonable to do so.[633] As stated above, this is the case whether or not there was a contractual restriction in the lease against assignment or subletting,[634] although it is more likely to be reasonable to make an order if there was a contractual restriction.

10.46 Particular points to note are:

- Case 6 can only be used when the protected tenancy has ended.

- The landlord's consent need not have been given *before* the assignment or subletting; Case 6 will not apply if he gave consent at any time before he begins a possession claim. Nor need consent be given expressly; it can be implied, but this will depend on all the circumstances.

- Case 6 may apply even if the situation has changed before the landlord starts his possession claim. For example, it can apply if:
    - the tenant sublets the whole property during the protected tenancy; but
    - the subtenancy comes to an end before the protected tenancy ends and the landlord starts a possession claim.[635]

    However, it is unlikely to be reasonable to order possession.

- If the tenant lawfully grants a subtenancy without landlord's consent (for example, because there is no covenant against subletting in the lease, so landlord's consent was not required), the subtenancy can continue even if the tenancy is ended by a court order on the basis of Case 6.[636] However, if the landlord has not accepted the subtenancy, he can also seek a court order terminating the subtenancy as well on the basis of Case 6; to rely on Case 6 against a subtenant, it is only necessary that the *tenant* has sublet, not that the subtenant has also assigned or sublet.[637]

Remedies for tenants

10.47 This section examines what the tenant's options are if the landlord wrongfully withholds consent. It assumes that the landlord has not given an express covenant not to withhold consent. In rare cases where such a covenant is given, the tenant will be entitled to use the remedies of damages or an injunction, which are described at 10.3 to 10.15.

Preliminary: does the landlord's failure help the tenant at all?

10.48 The tenant first needs to identify the type of covenant.

- *Absolute covenants:* If there is a complete prohibition, the landlord has no obligation to grant consent, or even respond to the tenant's application. The tenant therefore has no remedy.

- *Qualified covenants:* If there is a qualified restriction which the law does not treat as fully qualified (i.e. qualified covenants against changing use and limited other covenants explained in Appendix 2), again the tenant has no remedy.

  However, in most cases other than change of use involving structural alterations, the landlord may not demand a premium. If he does, the tenant will have limited remedies: see 10.55 to 10.63.

- *Fully qualified covenants:* If the covenant is fully qualified, the tenant will have a variety of remedies if the landlord:
  - unreasonably refuses consent;
  - grants consent subject to an unreasonable condition; or
  - delays in giving consent, which will be counted as an unreasonable withholding of consent.[638]

For an explanation of absolute, qualified and fully qualified covenants, see 1.6 to 1.15.

For details of when qualified covenants are made fully qualified and when a premium can lawfully be demanded, see 2.79 to 2.91, 3.35 to 3.51 and 4.50 to 4.64.

10.49 Paragraphs 10.50 to 10.79 refer to fully qualified covenants unless otherwise stated.

Go ahead anyway

10.50 The advantage of this remedy for tenants is that it is cheap, simple and quick. However, as explained at 10.59, it is very uncertain.

Fully qualified covenants

10.51 The tenant's primary option is to go ahead with whatever he applied to do. This applies to all types of fully qualified covenant, whether they relate to alienation, alterations or change of use. It also applies even where the tenant is able to claim damages, either because the LTA 1988 applies, as explained at 10.67 to 10.79, or because the landlord has expressly covenanted not to withhold consent unreasonably.[639]

10.52 If the tenant has covenanted that he will not do something without landlord's consent, such consent not to be unreasonably withheld, then if the tenant applies for consent and it is unreasonably withheld, the requirement for consent falls away.[640] This means that the tenant can do the restricted act without needing consent and will not be in breach of covenant.

10.53 The same applies if the landlord grants consent but subject to an unreasonable condition: the tenant is entitled to do the restricted act without satisfying the condition and without needing further consent, and will not be in breach of covenant.[641]

10.54 However, the tenant may only go ahead with the act for which he sought consent. If he modifies his plans, he must ask for consent again.

Example

A tenant seeks consent to assign his lease to a husband and wife. The landlord refuses consent on grounds that are plainly unreasonable. The tenant can go ahead with the assignment to the husband and wife. However, if the wife decides that she does not want to join in the transaction, the tenant cannot assign to the husband alone without asking for consent again.

Qualified covenants

10.55 The same principle can also apply to qualified covenants that are not fully qualified. In most cases other than a change of use that involves structural alterations, the landlord may not demand a premium. If the landlord does demand a premium, then the requirement for consent falls away. This means that the tenant can do the restricted act without paying the premium and without being in breach of covenant.[642]

For details of when qualified covenants are made fully qualified, and when a premium is forbidden, see 2.79 to 2.91, 3.35 to 3.51 and 4.50 to 4.64.

Absolute covenants

10.56 This principle does not apply to absolute covenants.

Assignments of secure tenancies by way of exchange: more certainty for tenants

10.57 As explained at 2.41, secure tenants can exchange their tenancies with the consent of their landlord. The landlord can only withhold consent to the assignment on certain grounds. If consent is not withheld on one of those grounds, it is deemed to have been given. There are strict rules on how quickly the landlord must respond, what information it must give in its response, and what conditions it can impose. These are explained at 6.30 to 6.31 and 7.185 to 7.186.

10.58 This means that if a secure tenant applies for consent to an exchange, he can be much more sure of his position than other tenants applying for consent to assign:

- If consent is given, subject to no conditions or conditions relating only to unpaid rent or breaches of covenant, he can go ahead provided he complies with the conditions.
- If consent is given subject to other conditions, he can go ahead without complying with the conditions. If consent is given subject to both permitted conditions (for example, one requiring rent arrears to be paid) and unlawful conditions (for example, one requiring the outgoing tenant to guarantee the incoming tenant's obligations), the tenant must comply with the permitted conditions but need not comply with the unlawful conditions.
- If the landlord has not replied within 42 days, the tenant can go ahead.
- If the landlord refuses consent without specifying the grounds for refusing, or without giving particulars of

them, the tenant can go ahead, although he should wait until 42 days have elapsed.

- If the landlord refuses consent on specified grounds, with particulars, the tenant should check the grounds in the notice against the grounds on which the landlord is entitled to rely (set out in Appendix 3 at pp. 296 to 299). If one or more of the grounds in the notice is a permitted ground, the tenant will *not* be entitled to go ahead. Otherwise, he will.

Declaration that consent has been unreasonably withheld

10.59 Although the tenant is entitled to go ahead without landlord's consent in the circumstances described above, this can be a very unsatisfactory solution.

If he is wrong that consent has been withheld unreasonably, or that a premium has been demanded unlawfully, then if he does go ahead, he will be in breach of covenant. This means that the landlord will be entitled to use the remedies described at 10.3 to 10.46. If the tenant is trying to assign the lease or grant a sublease, the prospective assignee or subtenant may be very reluctant to go ahead with the transaction without knowing for sure whether it is a breach of covenant or not.

10.60 To solve this problem, the tenant can ask the court for a declaration as to the position. If the tenant succeeds, the declaration will allow the tenant to go ahead safely, regardless of the landlord's consent.

10.61 Again, this remedy does not apply to:

- absolute covenants; or
- qualified covenants that are not made fully qualified and where a premium may be demanded (most commonly use covenants where structural alterations are involved).

For details of when qualified covenants are made fully qualified and when the landlord is prohibited from demanding a premium, see 2.79 to 2.91, 3.35 to 3.51 and 4.50 to 4.64.

10.62 Although it can be used for qualified covenants where an unlawful premium has been demanded (for example, if a landlord demands a premium for granting consent to a change of use not involving structural alterations), its main use is for fully qualified covenants.

10.63 This remedy does give a tenant certainty, but it can take a long time to get a declaration from the court, and it can also be expensive and complicated.

Who can seek the declaration?

10.64 The tenant can seek a declaration that consent has been unreasonably withheld and that he is entitled to do what he wants to do without further need for consent.

10.65 Alternatively, if the tenant has already gone ahead with an assignment or subletting, the assignee or subtenant can apply for a declaration.

10.66 Legal advice should be taken about what declaration to ask for, who to bring the claim against, and when it will be for the tenant rather than the landlord to prove that the landlord's withholding of consent was unreasonable.

Damages

10.67 A declaration that consent has been unreasonably withheld can also be an unsatisfactory remedy for a tenant, because of the length of time it takes to get a final court hearing. It is a particular problem in the case of assignments and sublettings, when the proposed assignee or subtenant might not be prepared to wait. If the assignee or subtenant withdraws from the transaction, the tenant may suffer loss.

10.68 Until 1988, if the landlord unreasonably refused consent to a tenant's application, the tenant usually had no right to damages, even if he suffered loss.[643] This is because the standard fully qualified covenant is not interpreted as a covenant by the landlord. If the landlord withholds consent unreasonably, he cannot stop the tenant from doing the act without his consent, but there were, until 1988, no other consequences.

10.69 The only exception to this rule was where the landlord gave an express covenant '*not to withhold consent unreasonably*', but it is very rare that a landlord does so.[644]

10.70 Since 1988, the position has been changed by the LTA 1988, but *only* for fully qualified covenants against assigning, subletting, charging and parting with possession. For all other types of covenant, the rule remains the same. Paragraphs 10.71 to 10.79 therefore only apply to those four types of alienation covenant. Unless the landlord has given an express covenant not to withhold consent unreasonably, there is still *no* right to damages in the case of:

- absolute covenants of any type;

- qualified covenants of any type which are not made fully qualified, regardless of whether a premium can be demanded or not;

- covenants against alterations;

- covenants against changing use;

- covenants against forms of alienation other than assigning, subletting, charging and parting with possession, such as making a declaration of trust; and

- covenants against assigning, subletting, charging and parting with possession in leases to which the LTA 1988 does not apply, such as secure tenancies.

For details on when the LTA 1988 applies and what duties it imposes, see 1.22 to 1.31.

See Appendix 2 for full details of when qualified covenants are made fully qualified.

10.71 If the landlord, or other relevant person, does not perform any of the duties imposed by the Act, the tenant has a right to claim damages against him. Legal advice should be taken about who can sue, who the claim should be brought against, and who must prove what.

Amount of damages

10.72 The tenant will usually be awarded a sum of money to compensate him for the loss caused by the landlord's breach of duty.

Examples

(1) A tenant applied for consent to sublet. The landlord withheld consent unreasonably and the underletting did not go ahead. The subtenant had been occupying the property under a licence but then moved out.

Had the underletting gone ahead, the tenant would have received rent and a contribution to insurance costs from the subtenant, and would not have had to pay rates. The landlord was ordered to pay damages to cover those items from the date when the subtenant gave up occupation under the licence to the date when it later went back into occupation under a new deal.[645]

(2) Another tenant applies for landlord's consent, having agreed to assign for a premium of £50,000. The landlord refuses consent unreasonably. It takes the tenant two years to find another assignee, and he has to pay rent, service charge, insurance premiums and rates for that period. He has already moved his business to other premises, so the building stands empty. When he does find another assignee, the market has dropped, and he can only get a premium of £30,000. He is likely to be able to claim the difference of £20,000 as well as the rent and other outgoings.

(3) A third tenant applies for consent to assign a residential flat for £200,000, having inherited a house which he intends to move to. The landlord unreasonably refuses consent. The tenant continues to live at the flat for three months, paying service charge of £100 a month. But when he sells three months later, he gets £210,000. He is unlikely to be awarded any damages at all.

10.73 Although the damages will usually be money to *compensate* the tenant for loss, it is also possible for *exemplary* damages to be awarded. These are damages which are awarded to punish the landlord's conduct. For example, exemplary damages of £25,000 were awarded in a case where the landlord had embarked on a deliberate policy to make a profit by delaying or refusing consent unreasonably.[646]

Causation of loss

10.74 The tenant must prove that its loss was *caused* by the landlord's breach of duty. If the transaction would not have gone ahead in any event, then the landlord's breach will not have caused the loss. At present, this is so even if the other reason why the transaction would have been lost is other wrongful behaviour by the landlord, but we doubt that this aspect of the principle will survive further consideration by the High Court or the Court of Appeal.

Example

A tenant wanted to assign its premises. It applied for consent to assign and to change use, both of which were subject to fully qualified restrictions. Consent was unreasonably refused for both. The tenant would be entitled, in principle, to claim damages for refusal of consent to assign, but not for change of use.

However, the landlord argued that even if it had granted consent to assign, the assignment would still not have gone ahead because of the refusal of consent to change use. There was no evidence from the assignee as to what it would have done in this situation. The judge concluded that he did not have enough evidence to justify the conclusion that, if the landlord had granted consent to assign, the outcome would have been different. Therefore the tenant's loss was not *caused* by the refusal of consent to assign, and no damages could be claimed.[647]

10.75 However, it need not be shown that the transaction would certainly have gone ahead; it can frequently be sufficient to show that the transaction would probably have gone ahead.

Example

When the attempt by the tenant in example (1) in 10.72 to sublet part of the premises failed because of the landlord's delay, a number of other points were still outstanding: the tenant and the subtenant had not finished negotiating the terms of an agreement for the tenant to contribute to the cost of installing a lift; the terms of the underlease had not yet been agreed in detail; and applications for change of use and alterations were still outstanding. The landlord argued that these outstanding matters meant that the delay in granting consent to sublet had not caused the loss; the subletting would not have completed anyway.

The judge held that in all probability, all of these matters would have been sorted out with sufficient speed if consent to the subletting had been granted within a reasonable time. The delay in granting consent to sublet had therefore caused the loss.[648]

When can the tenant sue?

10.76 The tenant will be able to bring a claim as soon as the landlord has broken one of his obligations and the tenant has suffered loss.

10.77 The first task, therefore, is to work out when the landlord breached his obligations. If the landlord promptly refuses consent, or grants consent subject to an unreasonable condition, it is easy to identify this: it is the date of the refusal or the date the condition was imposed. If the landlord does not respond at all or he delays, he is in breach as soon as a 'reasonable time' has elapsed. It can be difficult to identify what a reasonable time is, but some guidance is given at 6.26 to 6.29.

10.78 It may often be difficult for the tenant to work out what his loss is until some time later. For example, the tenant may lose an assignment and be unable to find another assignee for several years, and then only on worse terms. In those circumstances, it may be better for the tenant to give the landlord warning that he will be making the claim, but to defer the claim until he has a better idea of his losses.

10.79 However, the tenant should not delay too long. He must begin a claim within six years after the breach of duty, or his claim will be statute barred and will be dismissed without any investigation.[649]

Recovery of unlawful premium

See 1.19 for a summary of when the landlord may and may not ask for a premium, and 8.19 to 8.22 for details of what counts as a premium.

10.80 As set out at 2.79 to 2.91, 3.35 to 3.51 and 4.50 to 4.64, in many cases it is unlawful for a landlord to demand a premium in return for granting consent. As explained at

10.55, if the landlord does ask for a premium in those circumstances, the tenant is entitled to assign without further consent and without paying the premium.

10.81 Further, if the landlord grants consent on condition that the premium will be paid later, the landlord will be unable to enforce the agreement and sue for the unpaid premium.[650]

10.82 If the tenant actually does pay the premium, as the law presently stands, he will not be able to recover it.[651] However, it may be possible to argue that the law has moved on, and if an unlawful premium has been paid, legal advice should be sought.

Transfers of Rent Act statutory tenancies

A Rent Act tenancy is a type of private sector residential tenancy usually entered into before February 1989. For more details, see Appendix I.

10.83 As explained at 2.47, a Rent Act statutory tenant can agree to transfer his status to another person if his landlord agrees. However, anyone who requires the payment of money in return for entering into such an agreement can be fined, with certain exceptions such as apportioned outgoings.

10.84 Further, if money is demanded and paid it can be recovered. An order can be made either by the court that imposes the fine, or in a separate claim for recovery of the money. Alternatively, if the money was paid to the landlord by a person who pays him rent, the person can recover the money by withholding rent.[652]

Remedies against surveyors

10.85 Finally, a word of caution. As has been seen, if a landlord fails to respond promptly and reasonably to a tenant's application to assign, sublet, charge or part with possession, he may have to pay damages to the tenant. If his failure has been caused by his surveyor, the landlord is likely to look to the surveyor's insurers to make good his loss. It is therefore important for surveyors to be aware of the landlord's duties. In particular, surveyors must be prepared to act promptly, and to consider carefully any reasons for refusal or conditions for grant of consent.

10.86 Even for other types of application, it is important for surveyors to be aware of the landlord's position. As explained above, if a landlord may not withhold consent unreasonably and does so, or if he demands an unlawful premium, the tenant will be entitled to go ahead with his plans without any further need for consent. If this happens, the landlord may suffer loss.

Example

A tenant applies for consent to make an improvement. The works will diminish the letting value of the property. They will also have a small negative impact on the value of the landlord's neighbouring premises. The landlord could reasonably ask for approximately £5,000 compensation, and an undertaking from the tenant to reinstate at the end of the term.

The landlord's surveyor advises the landlord that this is a good opportunity to extract some money from the tenant, and with the landlord's authority responds to the tenant saying that consent will be granted in return for a premium of £10,000 on condition that the tenant undertakes to reinstate at the end of the term.

As explained at 3.45 to 3.48, it is impermissible to demand a premium in these circumstances. As explained at 10.50 to 10.54, this means that the tenant is entitled to go ahead with his improvements without the landlord's consent. He does not have to pay the £5,000 compensation. He may not even have to give the undertaking.

The landlord may well look to the surveyor to compensate him for the loss of the compensation, and for any diminution in the value of his property caused by the lack of the undertaking.

10.87 Tenants' surveyors must also be careful. For example, if a tenant asks a surveyor to make an application for consent on his behalf, and the surveyor fails to do so properly, the tenant's position may be adversely affected.

In particular, if the application is for consent to assign or sublet, then a failure to make an application in the proper form may deprive a tenant of a claim to damages. Alternatively, if a surveyor wrongly advises a tenant that an application is not necessary, the tenant may find himself in breach of covenant and the landlord may take action against him.

10.88 But with care and sense, surveyors can escape claims against them, and have satisfied clients.

Checklist for landlords: Enforcing rights

Use this checklist to decide what to do if a tenant has done something without the landlord's consent.

- Was the tenant's right to do the act restricted? See chapters 2, 3 and 4.
- If it was restricted, was the landlord's consent necessary? Could the landlord withhold consent unreasonably or demand a premium? See 1.6 to 1.21 and 1.33 to 1.52.
- If the landlord's consent was necessary, has he inadvertently granted consent? See 9.15 to 9.32. If so, the landlord will have no remedy.
- If the landlord could not withhold consent unreasonably or demand a premium, has he done so? See chapter 7. If so, the landlord will have no remedy.
- If the landlord has a remedy, does he want to terminate the lease? See 10.16 to 10.46 to see which regime applies. If the landlord does want to terminate the lease, has he waived the right to do so? See 10.21 to 10.26.
- If the landlord does not want to terminate the lease or cannot do so, see 10.3 to 10.15.

Checklist for tenants: Enforcing rights

Use this checklist to decide what to do if consent is withheld to the tenant's application.

- Does the tenant have a remedy at all? See 10.48.
- If the landlord has not responded to the tenant's application, has a reasonable time elapsed so that consent is treated as having been withheld? See 6.26 to 6.47.

- If consent has been withheld, was the withholding reasonable? If it has been granted subject to conditions, are the conditions unreasonable? See chapter 7.
- If consent has been unreasonably withheld, is the tenant (and the assignee or subtenant if appropriate) sufficiently confident to go ahead anyway?
- If not, does the tenant have the time and money to seek a declaration as to the position?
- Does the LTA 1988 apply? See 1.27 to 1.31. If it does, can the tenant prove that the landlord's conduct has caused him loss? See 10.72 to 10.75. If so, seek damages.

Appendix 1 – Key words and concepts explained

This appendix explains the meaning of:

- 'New' tenancy and 'old' tenancy;
- 'Lease' and 'licence';
- 'Assured' tenancy;
- 'Secure' tenancy;
- Rent Act 'protected' tenancy and 'statutory' tenancy; and
- Many other words and phrases common to landlords' consents.

It is intended to provide general guidance for surveyors, but unless the position is clear, it will often be necessary to take legal advice. For example, while we have given general guidance on residential tenancies which will be accurate in most situations, there are many exceptions.

New tenancies and old tenancies

A1.1   The LT(C)A 1995 created a dual regime for tenancies. As explained at 1.41 to 1.46, 'new' tenancies can prescribe circumstances in which it will be reasonable for a landlord to withhold consent to an assignment. 'Old' tenancies can lay down pre-conditions to making an application for consent, but cannot prescribe when it will be reasonable to refuse the application. The distinction also affects whether tenants and their guarantors continue to be liable on the covenants in the lease after they have assigned the lease on, but that is outside the scope of this book.

A1.2   The primary rule is that a tenancy is a new tenancy if it was granted on or after 1 January 1996.[653]

A1.3   However, a tenancy granted on or after 1 January 1996 is not a new tenancy if the obligation to grant it pre-dated 1 January 1996. So a tenancy will not be a new tenancy if it was granted pursuant to:

- a pre-1996 contract;

- a pre-1996 option or right of first refusal, regardless of when the option is exercised;
- a pre-1996 court order; or
- s. 19 of the 1995 Act (which relates to 'overriding' leases, which a previous tenant can take when he has to pay arrears owed by a subsequent tenant), and the original tenancy was not a new tenancy.

A1.4   A surrender of a tenancy and immediate grant of a new one after 1 January 1996 will create a new tenancy unless the obligation to enter into the transaction arose before 1996.[654] This can happen accidentally: if a landlord and a tenant agree to vary the lease by extending the length of the term or increasing the size of the demised premises, this will take effect as a surrender and re-grant unless it is carefully structured. It is unlikely that any other variations will have this effect.[655]

A1.5   Whether a fresh lease taken up by a tenant's surety following disclaimer or forfeiture will be a new tenancy depends on when the agreement by the surety to take the lease was made. If it was made before 1996, it will not be a new tenancy.

A1.6   Tenancies that are not new tenancies are usually referred to as 'old' tenancies.

Leases and licences

A1.7   It can be very difficult to tell the difference between an occupational lease and a licence. But leases and licences can have very different effects in law.

A1.8   A lease creates an 'estate in land', whereas a licence is merely a contractual right to be on the premises. A lease must usually be for a defined period and rent must usually be payable (although there can be exceptions to both rules), but the key difference between a lease and a licence is 'exclusive possession'.

A1.9   There is no clear definition of exclusive possession. It is not the same as occupation. A licensee can be in exclusive occupation without having exclusive possession and therefore having a lease.

A1.10 However, the following points should be noted:

- It is the substance of the arrangement that is important. Calling an arrangement a licence will not make it a licence if it is actually a lease.
- An express reservation of a right for the 'landlord' to enter the property, for example to carry out repairs, will usually point to a tenancy.
- If the 'landlord' provides services in the demised property, such as cleaning, there will usually be a licence.
- If there is no intention to create a legal relationship at all, such as when the 'landlord' allows the 'tenant' into the property as a temporary arrangement out of charity, there will usually be a licence.
- If the 'landlord' retains overall control of the property (for example, if he can require the 'tenant' to move to a different unit or require him to share the property with other occupiers), there will usually be a licence.

Assured tenancies

A1.11 Most private sector residential tenancies granted since 15 January 1989 are known as 'assured' tenancies. These impose restrictions on the landlord's right to recover possession at the end of the tenancy. Many are 'assured shorthold' tenancies, a subset of assured tenancies which give the tenant less security from eviction at the end of the tenancy. Assured tenancies must fulfil certain criteria:

- The tenant must be an individual (i.e. not a company), or if the tenancy is a joint tenancy, each of the tenants must be an individual.
- The property must be a dwelling-house let as a separate dwelling which the individual occupies as his only or principal home.

A1.12 However, many tenancies that fulfil these criteria are not assured. The most important exceptions are:

- tenancies where the rent is more than £25,000 a year;
- tenancies at no rent, or a low rent (this will exclude most long residential tenancies); and
- tenancies with a resident landlord.

A1.13 Assured tenancies can be for a fixed term, or they can be periodic, i.e. the tenancy continues from week to week or month to month unless either the landlord or the tenant takes action to bring it to an end in a permitted way. Whichever type of tenancy it is, the tenant cannot be required to leave unless the landlord obtains a court order.

Secure tenancies

A1.14 Most residential tenancies granted by public sector bodies are known as 'secure' tenancies. Secure tenancies must fulfil certain criteria:

- The tenant must be an individual (i.e. not a company), or if the tenancy is a joint tenancy, each of the tenants must be an individual.
- The property must be a dwelling-house (often a council flat), which the individual occupies as his only or principal home.
- The landlord must be within a set group of organisations, but in practice is often a local authority or housing trust.

A1.15 Again, there are a number of exceptions.

A1.16 If the tenancy is secure, the tenant has additional rights and duties. Most importantly, there are statutory restrictions on the landlord's ability to recover possession of the premises and, in certain circumstances, family members can 'succeed' to the secure tenancy when the tenant dies. The family member then retains the benefits of the tenancy.

A1.17 Secure tenancies can be for a fixed term, or they can be periodic, i.e. the tenancy continues from week to week or month to month unless either the landlord or the tenant takes action to bring it to an end in a permitted way. Whichever type of tenancy it is, it cannot be ended against the tenant's will unless the landlord obtains a court order.

Rent Act tenancies

A1.18 The RA 1977 created a special type of tenancy for residential tenants, which carried a great deal of protection from eviction. Most private sector residential tenancies granted before 15 January 1989 will be governed by this Act. Again, there are a number of exceptions.

A1.19 As from 15 January 1989, Rent Act tenancies were replaced by assured tenancies. But this only applied to new lettings from that date; existing tenancies continue to be Rent Act tenancies. In certain circumstances, tenancies created after 15 January 1989 can still be Rent Act tenancies.[656] The most important exceptions are:

- If the tenancy was created after 15 January 1989 pursuant to a contract entered into before 15 January 1989, the tenancy will be governed by the RA 1977.
- If a Rent Act protected or statutory tenant takes a new tenancy from the same landlord (whether of the same property or a different one), the new tenancy will be governed by the RA 1977.

Protected tenancies

A1.20 Rent Act tenancies can be for a fixed term, or they can be periodic, i.e. the tenancy continues from week to week or month to month unless either the landlord or the tenant takes action to bring it to an end in a permitted way. So long as the contractual tenancy continues to exist, the tenancy is 'protected'.

A1.21 The landlord cannot seek a possession order until the protected tenancy is terminated, for example by expiry of a fixed term, by service of a notice to quit for a periodic tenancy, or by the landlord forfeiting the lease if the tenant is in breach of covenant.

Statutory tenancies

A1.22 When a protected tenancy is terminated, if, and so long as, the tenant continues to live at the property he cannot be evicted. Instead, he can stay at the property under a 'statutory' tenancy. A statutory tenancy cannot be terminated by the landlord unless he obtains a court order for possession. He can only obtain a court order in certain circumstances. The most common is that the tenant has failed to pay the rent or is in breach of a covenant in the tenancy agreement.

Glossary of terms

Many of the words and phrases used in connection with landlords' consents have particular meanings. This glossary gives a general guide to the meaning of some of the words and phrases most commonly encountered. Where required, specific legal advice should be obtained as to their precise meanings.

Word or phrase	Meaning
Absolute covenant	A *covenant* that completely prohibits a certain act or thing. See also 1.8 to 1.9.
Address for service	The location where a person or company may be served with *notices* or other legal documents.
Agreement for lease	A legal agreement that one party will grant, and another party will take, a *lease*.
Air space	The air above a property or area of land, which may or may not be included in a *demise* of the property.
Alienation	The process of granting or parting with an interest in a property. See also chapter 2.
Alteration	An alteration to the form or structure of a property. See also chapter 3.
Assent	A transfer of property from personal representatives to a beneficiary.
Assign	To transfer rights to property.
Assignee	The person to whom an *assignment* is made.
Assignment	A transfer of rights to property.
Assignor	The person who makes an *assignment*.
Assurance	Another word for an *assignment* or transfer.
Assured shorthold	See A1.11 to A1.13.
Assured tenancy	See A1.11 to A1.13.
Authorised Guarantee Agreement ('AGA')	A type of *guarantee*, the terms of which are controlled by law, given by a *tenant* when assigning a *lease*. See also 7.133 to 7.137.
Bankruptcy	The formal process of distributing the assets of an insolvent individual.
Breach	A failure to comply with an obligation.
Charge	To charge is to give property as security for a debt. A charge is the security itself.
Claimant	The name given to a person making a court claim.
Claim Form	The form used to begin a court claim.

Word or phrase	Meaning
Collateral agreement	A separate agreement, ancillary to a main contract or lease. See also *Side letter*.
Commencement date	The beginning of the *term* of a *lease*. Whether the commencement date itself is included in the *term* depends on the precise words used.
Common parts	This phrase usually refers to the areas of a property available for use by any of the *occupiers* (for example, communal hallways or grounds).
Consent	Agreement or permission to do a certain act.
Co-owner	A person who owns property jointly with one or more others.
Costs	Expenses. For example, those incurred dealing with an application for *consent* or in *litigation*. See also chapter 8.
Covenant	A contractual obligation (strictly one given in a *deed*).
Damages	Compensation awarded by a court.
Declaration	A statement of a person's rights or obligations given by a court.
Deed	A type of agreement or formal *undertaking* which satisfies certain conditions. The conditions are, broadly speaking, that the document must make clear on its face that it is intended to be a deed and that it must be signed and witnessed (by an individual) or executed (by a company) and delivered.[657]
Defendant	A person against whom a court claim is made.
Demise	The property included in a *lease* (or the act of granting a *lease*).
Determination	Synonym of *termination*.
Disposal	The process of granting or parting with an interest in a property.
Disposition	See *Disposal*.
Dwelling	A building (or part of a building) intended to be used for living in.
Estoppel	A legal concept by which a person is prevented, or 'estopped', from relying on their legal rights.
Execution	The process by which a deed is validated, for example by signing or sealing.
Express term	A term in a contract or *lease* which the parties have expressly agreed to. See also *Implied term*.
Fine	A *premium* or sum of money (often required as a condition of giving *consent* or of granting a *lease*). See also 8.19 to 8.22.

Word or phrase	Meaning
Forfeiture	Termination of a *lease* by the landlord because of default by the *tenant*.
Freehold	A form of ownership whereby, for practical purposes, the property belongs to the owner absolutely.
Fully qualified covenant	A *covenant* that prohibits an act without *consent*, where the *consent* cannot be unreasonably withheld. See also 1.13 to 1.14.
Guarantee	A promise by a second person to fulfil an obligation if the first person fails to do so.
Guarantor	A person who gives a *guarantee*.
Holding over	The act of remaining in *possession* of property after the formal right to do so has ended.
Hybrid covenant	A *covenant* that has features of both absolute and qualified *covenants*. See also 1.15.
Implied term	A term that is deemed to appear in a contract or *lease* even if not expressly included. See also *Express term*.
Improvement	This word has different meanings depending on the context. It often means an *alteration* that has the effect of improving *premises* from the *tenant's* point of view. See also 3.43 and 3.79.
Injunction	A court order requiring a person to do (or refrain from doing) something.
Insolvency	The inability to pay debts as they fall due.
'Keep open' covenant	A *covenant* requiring the *tenant* to keep his property open for business. See 4.9 to 4.12 and 10.13.
Landlord	The person by whom a *lease* is granted (or the person who currently owns the *reversion* to the *lease*).
Lands Tribunal	A court with specific jurisdiction over certain types of dispute, usually involving *leases* or land, and which also hears appeals from the *Leasehold Valuation Tribunal*.
Lease	See A1.7 to A1.10.
Leasehold	The form of land ownership deriving from a *lease*.
Leasehold Valuation Tribunal	A court with specific jurisdiction over certain types of dispute, usually involving *leases* or land.
Lessee	Synonym of *tenant*.
Lessor	Synonym of *landlord*.
Letting	Granting a *lease*; sometimes the lease itself.

Word or phrase	Meaning
Licence	A right or permission to do something (for example to assign a lease or to occupy property). See A1.7 to A1.10.
Limitation	The requirement for legal claims to be started within a certain time.
Liquidation	The formal process of winding up a company (often because of *insolvency*).
Liquidator	The person responsible for winding up a company in *liquidation*.
Litigation	The process of resolving a dispute through the courts.
Mixed use	The use of *premises* (or parts of *premises*) for different purposes (often part business use and part residential).
Mortgage	Synonym of *charge*.
Mortgagee	The recipient of the security given by the *mortgage* (usually a bank or other lender).
Mortgagor	The person who gives the *mortgage* – the owner of the property.
New tenancy	See A1.1 to A1.6.
Notice	Formal notification.
Occupation	Making use of *premises* (but not necessarily having *possession* of the *premises*).
Occupier	A person who is in *occupation*.
Old tenancy	See A1.1 to A1.6.
Periodic tenancy	A type of *tenancy* that continues indefinitely from one period, such as a month or a year, to another until terminated by *notice*.
Permit or suffer	Words that are commonly used to widen the ambit of a *covenant*, to include (broadly) not only things done by the covenantor personally but also things that he allows or does not prevent others from doing.
Possession	A legal concept relating to the physical control of property. Contrast with *occupation*. See also A1.7 to A1.10.
Pre-condition	A requirement that certain circumstances exist before an event occurs. For examples in the context of landlords' consents, see 1.36 to 1.40 and 1.15.
Premises	Land or buildings (possibly including the *air space* above them). The word is usually used to describe the extent of the *demise* in a *lease*.
Premium	A sum of money or something of value.

Word or phrase	Meaning
Protected tenancy	See A1.18 to A1.22.
Proviso	A clause (or part of a clause) that qualifies the meaning of an agreement.
Qualified covenant	A *covenant* that prohibits an act without *consent,* but where there is no *proviso* that *consent* is not to be unreasonably withheld. Contrast with *fully qualified covenant.* See also 1.10 to 1.12.
Reinstatement	The act of removing or making good *alterations.*
Remedy	A method of enforcing legal rights or seeking redress.
Rent	The sum payable by a *tenant* to a *landlord* for *possession* of property.
Reverse premium	A *premium* paid in order to induce someone to take on an obligation (for example to acquire a *lease*).
Reversion	The right to have property back after the end of a *lease.* A *landlord* is said to be the owner of the *reversion* to the *lease.*
Secure tenancy	See A1.14 to A1.17.
Serving notice	The act of giving a *notice* to someone.
Sharing occupation (or possession)	The act of allowing another person to jointly occupy (or possess) property.
Side letter	A short form of *collateral agreement*, often used to record minor *variations* to a *lease.*
Statutory tenancy	See A1.18 to A1.22.
Structural alterations	Broadly, an *alteration* that affects a building's structure or which is itself structural.
Sublease	A *lease* where the *landlord's* interest is also *leasehold.*
Suffer	See *Permit or suffer.*
Surrender	The act of ending a *lease* by transferring it back to the *landlord.*
Tenancy	Synonym of *lease.*
Tenant	The person to whom a *lease* is granted (or the person currently entitled to the rights granted by the *lease*).
Term	The duration of a *lease.* Also a clause or provision in a *lease* or agreement.
Termination	The ending of a *lease* or agreement.
Tribunal	A forum for resolving disputes. The term usually indicates a body outside the usual court system.
Underlease	Synonym of *sublease.*

Word or phrase	Meaning
Undertaking	A legally enforceable promise.
User	In the context of landlords' consents, this indicates the type of use permitted under a *lease*.
Variation (of a lease)	A formal change to the pre-existing terms of a *lease*.
Waste	A legal concept involving some destruction, or a change in the character, of a property. See also 3.28 to 3.30.

Appendix 2 – Transformation of qualified covenants

As summarised at 1.19, many qualified covenants are automatically transformed by law. Either they are rendered fully qualified, or landlords are prevented from demanding a premium in return for giving consent.

This appendix explains precisely which covenants are transformed. See 2.79 to 2.91, 3.35 to 3.51 and 4.50 to 4.64 for details of the effect of the transformations.

It deals only with alienation, alterations and change of use covenants; it does not deal with other covenants where consent may be required, such as covenants not to apply for planning permission without consent, or covenants to insure in an insurance office approved by the landlord. Such covenants are rarely, if ever, transformed.

Alienation

Prohibition on premiums

> For details of what constitutes a premium, see 8.19 to 8.22.

A2.1 The LPA 1925, s. 144 prohibits the landlord from demanding a premium unless the lease expressly permits it.

Leases affected

A2.2 The section applies to all leases, even if they were entered into before the LPA 1925 came into force.[658]

Covenants affected

A2.3 The section applies to all qualified and fully qualified covenants against:

279

- assigning;
- underletting;
- parting with possession; or
- disposing of premises.

A2.4 It certainly applies to assignments etc. of the whole of the demised premises. It is unclear whether it applies to assignments etc. of part only.

Qualified restrictions made fully qualified

A2.5 The LTA 1927, s. 19(1)(a) transforms qualified alienation covenants into fully qualified ones.

Leases affected

A2.6 The starting point is that the transformation applies to all leases and agreements for lease,[659] even if they were entered into before the LTA 1927 came into force.

A2.7 However, it does not apply to:

- assured tenancies where the qualified covenant is implied (see 2.56 to 2.61);
- qualified covenants against subletting in certain leases extended under the *Leasehold Reform Act 1967* where the covenant is expressed to be entered into to give effect to s. 30 of that Act;[660]
- leases of agricultural holdings (usually agricultural tenancies granted before 1 September 1995);[661]
- farm business tenancies (usually agricultural tenancies granted on or after 1 September 1995);[662] or
- leases granted to contractors in respect of the running of prisons,[663] secure training centres[664] or removal centres.[665]

Covenants affected

A2.8 The transformation applies to all qualified covenants against:

- assigning;
- subletting;
- charging; or
- parting with possession.

A2.9 It applies to assignments etc. of the whole or part of the demised premises.

A2.10 It also applies to fully qualified covenants against assignment etc. If the transformation of the covenant is more favourable to the tenant than the contractual provisions of the lease, the tenant will be able to rely on the transformation.[666]

A2.11 The same leases and covenants are affected by s. 19(1)(b). This relieves the tenants of some long building leases from the need to seek consent at all. It is explained at 2.94 to 2.99.

Alterations

Prohibition on premiums

A2.12 There is no specific prohibition on landlords demanding a premium for consent to alterations.

Qualified restrictions made fully qualified

A2.13 The LTA 1927, s. 19(2) transforms qualified alterations covenants into fully qualified ones.

Leases affected

A2.14 The starting point is that the transformation applies to all leases and agreements for lease,[667] even if they were entered into before the LTA 1927 came into force.

A2.15 However, it does not apply to:

- leases of agricultural holdings (usually agricultural tenancies granted before 1 September 1995);[668]
- farm business tenancies (usually agricultural tenancies granted on or after 1 September 1995);[669]
- leases granted to contractors in respect of the running of prisons,[670] secure training centres[671] or removal centres;[672]
- mining leases;[673] or
- Rent Act or secure tenancies (although these have similar provisions as explained at 3.31 to 3.34).

A Rent Act tenancy is a type of private sector residential tenancy usually granted before 1989. A secure tenancy is a type of public sector residential tenancy. See Appendix 1 for more details.

Covenants affected

A2.16 The transformation applies to all qualified covenants against improvements. For details of what constitutes an 'improvement', see 3.43.

A2.17 It also applies to fully qualified covenants against improvements. If the transformation of the covenant is more favourable to the tenant than the contractual provisions of the lease, the tenant will be able to rely on the transformation.[674]

Change of use

Prohibition on premiums

A2.18 The LTA 1927, s. 19(3) prohibits the landlord from demanding a premium where there is a qualified use covenant. The lease cannot override this.

Leases affected

A2.19 The starting point is that the prohibition applies to all leases and agreements for lease,[675] even if they were entered into before the LTA 1927 came into force.

A2.20 However, it does not apply to:
- leases of agricultural holdings (usually agricultural tenancies granted before 1 September 1995);[676]
- farm business tenancies (usually agricultural tenancies granted on or after 1 September 1995);[677]
- leases granted to contractors in respect of the running of prisons,[678] secure training centres[679] or removal centres;[680] or
- mining leases.[681]

Covenants affected

A2.21 The subsection applies to all qualified and fully qualified covenants against changing use where the change does not involve any structural alteration of the premises.

Qualified restrictions made fully qualified

A2.22 Qualified covenants against changing use are not made fully qualified.

Appendix 3 – Selected legislation

Landlord and Tenant Act 1988

1 Qualified duty to consent to assigning, underletting etc of premises

(1) This section applies in any case where—
    (a) a tenancy includes a covenant on the part of the tenant not to enter into one or more of the following transactions, that is—
        (i) assigning,
        (ii) underletting,
        (iii) charging, or
        (iv) parting with the possession of,

    the premises comprised in the tenancy or any part of the premises without the consent of the landlord or some other person, but
    (b) the covenant is subject to the qualification that the consent is not to be unreasonably withheld (whether or not it is also subject to any other qualification).

(2) In this section and section 2 of this Act—
    (a) references to a proposed transaction are to any assignment, underletting, charging or parting with possession to which the covenant relates, and
    (b) references to the person who may consent to such a transaction are to the person who under the covenant may consent to the tenant entering into the proposed transaction.

(3) Where there is served on the person who may consent to a proposed transaction a written application by the tenant for consent to the transaction, he owes a duty to the tenant within a reasonable time—
    (a) to give consent, except in a case where it is reasonable not to give consent,
    (b) to serve on the tenant written notice of his decision whether or not to give consent specifying in addition—
        (i) if the consent is given subject to conditions, the conditions,

283

(ii) if the consent is withheld, the reasons for withholding it.

(4) Giving consent subject to any condition that is not a reasonable condition does not satisfy the duty under subsection (3)(a) above.

(5) For the purposes of this Act it is reasonable for a person not to give consent to a proposed transaction only in a case where, if he withheld consent and the tenant completed the transaction, the tenant would be in breach of a covenant.

(6) It is for the person who owed any duty under subsection (3) above—
   (a) if he gave consent and the question arises whether he gave it within a reasonable time, to show that he did,
   (b) if he gave consent subject to any condition and the question arises whether the condition was a reasonable condition, to show that it was,
   (c) if he did not give consent and the question arises whether it was reasonable for him not to do so, to show that it was reasonable,

and, if the question arises whether he served notice under that subsection within a reasonable time, to show that he did.

2 Duty to pass on applications

(1) If, in a case where section 1 of this Act applies, any person receives a written application by the tenant for consent to a proposed transaction and that person—
   (a) is a person who may consent to the transaction or (though not such a person) is the landlord, and
   (b) believes that another person, other than a person who he believes has received the application or a copy of it, is a person who may consent to the transaction,

he owes a duty to the tenant (whether or not he owes him any duty under section 1 of this Act) to take such steps as are reasonable to secure the receipt within a reasonable time by the other person of a copy of the application.

(2) The reference in section 1(3) of this Act to the service of an application on a person who may consent to a proposed transaction includes a reference to the receipt by him of an application or a copy of an application (whether it is for his consent or that of another).

3 Qualified duty to approve consent by another

(1) This section applies in any case where—
    (a) a tenancy includes a covenant on the part of the tenant not without the approval of the landlord to consent to the sub-tenant—
        (i)   assigning,
        (ii) underletting,
        (iii) charging, or
        (iv) parting with the possession of,
    the premises comprised in the sub-tenancy or any part of the premises, but
    (b) the covenant is subject to the qualification that the approval is not to be unreasonably withheld (whether or not it is also subject to any other qualification).

(2) Where there is served on the landlord a written application by the tenant for approval or a copy of a written application to the tenant by the sub-tenant for consent to a transaction to which the covenant relates the landlord owes a duty to the sub-tenant within a reasonable time—
    (a) to give approval, except in a case where it is reasonable not to give approval,
    (b) to serve on the tenant and the sub-tenant written notice of his decision whether or not to give approval specifying in addition—
        (i) if approval is given subject to conditions, the conditions,
        (ii) if approval is withheld, the reasons for withholding it.

(3) Giving approval subject to any condition that is not a reasonable condition does not satisfy the duty under subsection (2)(a) above.

(4) For the purposes of this section it is reasonable for the landlord not to give approval only in a case where, if he withheld approval and the tenant gave his consent, the tenant would be in breach of covenant.

(5) It is for a landlord who owed any duty under subsection (2) above—
    (a) if he gave approval and the question arises whether he gave it within a reasonable time, to show that he did,

  (b) if he gave approval subject to any condition and the question arises whether the condition was a reasonable condition, to show that it was,

  (c) if he did not give approval and the question arises whether it was reasonable for him not to do so, to show that it was reasonable,

and, if the question arises whether he served notice under that subsection within a reasonable time, to show that he did.

4 Breach of duty

A claim that a person has broken any duty under this Act may be made the subject of civil proceedings in like manner as any other claim in tort for breach of statutory duty.

5 Interpretation

(1) In this Act—

  "covenant" includes condition and agreement,

  "consent" includes licence,

  "landlord" includes any superior landlord from whom the tenant's immediate landlord directly or indirectly holds,

  "tenancy", subject to subsection (3) below, means any lease or other tenancy (whether made before or after the coming into force of this Act) and includes—

   (a) a sub-tenancy, and

   (b) an agreement for a tenancy

  and references in this Act to the landlord and to the tenant are to be interpreted accordingly, and

  "tenant", where the tenancy is affected by a mortgage (within the meaning of the Law of Property Act 1925) and the mortgagee proposes to exercise his statutory or express power of sale, includes the mortgagee.

(2) An application or notice is to be treated as served for the purposes of this Act if—

  (a) served in any manner provided in the tenancy, and

  (b) in respect of any matter for which the tenancy makes no provision, served in any manner provided by section 23 of the Landlord and Tenant Act 1927.

(3) This Act does not apply to a secure tenancy (defined in section 79 of the Housing Act 1985) [or to an introductory tenancy (within the meaning of Chapter I of Part V of the Housing Act 1996)].

(4) This Act applies only to applications for consent or approval served after its coming into force.

6 Application to Crown

This Act binds the Crown; but as regards the Crown's liability in tort shall not bind the Crown further than the Crown is made liable in tort by the Crown Proceedings Act 1947.

7 Short title, commencement and extent

(1) This Act may be cited as the Landlord and Tenant Act 1988.

(2) This Act shall come into force at the end of the period of two months beginning with the day on which it is passed.

(3) This Act extends to England and Wales only.

Landlord and Tenant Act 1927

1 Tenant's right to compensation for improvements

(1) Subject to the provisions of this Part of this Act, a tenant of a holding to which this Part of this Act applies shall, if a claim for the purpose is made in the prescribed manner [and within the time limited by section forty-seven of the Landlord and Tenant Act 1954], be entitled, at the termination of the tenancy, on quitting his holding, to be paid by his landlord compensation in respect of any improvement (including the erection of any building) on his holding made by him or his predecessors in title, not being a trade or other fixture which the tenant is by law entitled to remove, which at the termination of the tenancy adds to the letting value of the holding:

Provided that the sum to be paid as compensation for any improvement shall not exceed—
   (a) the net addition to the value of the holding as a whole which may be determined to be the direct result of the improvement; or
   (b) the reasonable cost of carrying out the improvement at the termination of the tenancy, subject to a deduction of an amount equal to the cost (if any) of putting the works constituting the improvement into a reasonable state of repair, except so far as such cost

is covered by the liability of the tenant under any covenant or agreement as to the repair of the premises.

(2) In determining the amount of such net addition as aforesaid, regard shall be had to the purposes for which it is intended that the premises shall be used after the termination of the tenancy, and if it is shown that it is intended to demolish or to make structural alterations in the premises or any part thereof or to use the premises for a different purpose, regard shall be had to the effect of such demolition, alteration or change of user on the additional value attributable to the improvement, and to the length of time likely to elapse between the termination of the tenancy and the demolition, alteration or change of user.

(3) In the absence of agreement between the parties, all questions as to the right to compensation under this section, or as to the amount thereof, shall be determined by the tribunal hereinafter mentioned, and if the tribunal determines that, on account of the intention to demolish or alter or to change the user of the premises, no compensation or a reduced amount of compensation shall be paid, the tribunal may authorise a further application for compensation to be made by the tenant if effect is not given to the intention within such time as may be fixed by the tribunal.

3 Landlord's right to object

(1) Where a tenant of a holding to which this Part of this Act applies proposes to make an improvement on his holding, he shall serve on his landlord notice of his intention to make such improvement, together with a specification and plan showing the proposed improvement and the part of the existing premises affected thereby, and if the landlord, within three months after the service of the notice, serves on the tenant notice of objection, the tenant may, in the prescribed manner, apply to the tribunal, and the tribunal may, after ascertaining that notice of such intention has been served upon any superior landlords interested and after giving such persons an opportunity of being heard, if satisfied that the improvement—

(a) is of such a nature as to be calculated to add to the letting value of the holding at the termination of the tenancy; and

(b) is reasonable and suitable to the character thereof; and

(c) will not diminish the value of any other property belonging to the same landlord, or to any superior landlord from whom the immediate landlord of the tenant directly or indirectly holds;

and after making such modifications (if any) in the specification or plan as the tribunal thinks fit, or imposing such other conditions as the tribunal may think reasonable, certify in the prescribed manner that the improvement is a proper improvement:

Provided that, if the landlord proves that he has offered to execute the improvement himself in consideration of a reasonable increase of rent, or of such increase of rent as the tribunal may determine, the tribunal shall not give a certificate under this section unless it is subsequently shown to the satisfaction of the tribunal that the landlord has failed to carry out his undertaking.

(2) In considering whether the improvement is reasonable and suitable to the character of the holding, the tribunal shall have regard to any evidence brought before it by the landlord or any superior landlord (but not any other person) that the improvement is calculated to injure the amenity or convenience of the neighbourhood.

(3) The tenant shall, at the request of any superior landlord or at the request of the tribunal, supply such copies of the plans and specifications of the proposed improvement as may be required.

(4) Where no such notice of objection as aforesaid to a proposed improvement has been served within the time allowed by this section, or where the tribunal has certified an improvement to be a proper improvement, it shall be lawful for the tenant as against the immediate and any superior landlord to execute the improvement according to the plan and specification served on the landlord, or according to such plan and specification as modified by the tribunal or by agreement between the tenant and the landlord or landlords affected, anything in any lease of the premises to the contrary notwithstanding:

Provided that nothing in this subsection shall authorise a tenant to execute an improvement in contravention of any restriction created or imposed

(a) for naval, military or air force purposes;

(b) for civil aviation purposes under the powers of the Air Navigation Act 1920;

(c) for securing any rights of the public over the foreshore or bed of the sea.

(5) A tenant shall not be entitled to claim compensation under this Part of this Act in respect of any improvement unless he has, or his predecessors in title have, served notice of the proposal to make the improvement under this section, and (in case the landlord has served notice of objection thereto) the improvement has been certified by the tribunal to be a proper improvement and the tenant has complied with the conditions, if any, imposed by the tribunal, nor unless the improvement is completed within such time after the service on the landlord of the notice of the proposed improvement as may be agreed between the tenant and the landlord or may be fixed by the tribunal, and where proceedings have been taken before the tribunal, the tribunal may defer making any order as to costs until the expiration of the time so fixed for the completion of the improvement.

(6) Where a tenant has executed an improvement of which he has served notice in accordance with this section and with respect to which either no notice of objection has been served by the landlord or a certificate that it is a proper improvement has been obtained from the tribunal, the tenant may require the landlord to furnish to him a certificate that the improvement has been duly executed; and if the landlord refuses or fails within one month after the service of the requisition to do so, the tenant may apply to the tribunal who, if satisfied that the improvement has been duly executed, shall give a certificate to that effect.

Where the landlord furnishes such a certificate, the tenant shall be liable to pay any reasonable expenses incurred for the purpose by the landlord, and if any question arises as to the reasonableness of such expenses, it shall be determined by the tribunal.

19 Provisions as to covenants not to assign, etc, without licence or consent

(1) In all leases whether made before or after the commencement of this Act containing a covenant condition or agreement against assigning, underletting, charging or parting with the possession of demised premises or any part thereof without licence or consent, such covenant condition or agreement shall, notwithstanding any express provision to the contrary, be deemed to be subject—

  (a) to a proviso to the effect that such licence or consent is not to be unreasonably withheld, but this proviso does not preclude the right of the landlord to require payment of a reasonable sum in respect of any legal or other expenses incurred in connection with such licence or consent; and

  (b) (if the lease is for more than forty years, and is made in consideration wholly or partially of the erection, or the substantial improvement, addition or alteration of buildings, and the lessor is not a Government department or local or public authority, or a statutory or public utility company) to a proviso to the effect that in the case of any assignment, under-letting, charging or parting with the possession (whether by the holders of the lease or any under-tenant whether immediate or not) effected more than seven years before the end of the term no consent or licence shall be required, if notice in writing of the transaction is given to the lessor within six months after the transaction is effected.

[(1A) Where the landlord and the tenant under a qualifying lease have entered into an agreement specifying for the purposes of this subsection—

  (a) any circumstances in which the landlord may withhold his licence or consent to an assignment of the demised premises or any part of them, or

  (b) any conditions subject to which any such licence or consent may be granted,

  then the landlord—

  (i) shall not be regarded as unreasonably withholding his licence or consent to any such assignment if he withholds it on the ground (and it is the case) that any such circumstances exist, and

(II) if he gives any such licence or consent subject to any such conditions, shall not be regarded as giving it subject to unreasonable conditions,

and section 1 of the Landlord and Tenant Act 1988 (qualified duty to consent to assignment etc) shall have effect subject to the provisions of this subsection.

(1B) Subsection (1A) of this section applies to such an agreement as is mentioned in that subsection—

(a) whether it is contained in the lease or not, and

(b) whether it is made at the time when the lease is granted or at any other time falling before the application for the landlord's licence or consent is made.

(1C) Subsection (1A) shall not, however, apply to any such agreement to the extent that any circumstances or conditions specified in it are framed by reference to any matter falling to be determined by the landlord or by any other person for the purposes of the agreement, unless under the terms of the agreement—

(a) that person's power to determine that matter is required to be exercised reasonably, or

(b) the tenant is given an unrestricted right to have any such determination reviewed by a person independent of both landlord and tenant whose identity is ascertainable by reference to the agreement,

and in the latter case the agreement provides for the determination made by any such independent person on the review to be conclusive as to the matter in question.

(1D) In its application to a qualifying lease, subsection (1)(b) of this section shall not have effect in relation to any assignment of the lease.

(1E) In subsections (1A) and (1D) of this section—

(a) "qualifying lease" means any lease which is a new tenancy for the purposes of section 1 of the Landlord and Tenant (Covenants) Act 1995 other than a residential lease, namely a lease by which a building or part of a building is let wholly or mainly as a single private residence; and

(b) references to assignment include parting with possession on assignment.]

(2) In all leases whether made before or after the commencement of this Act containing a covenant condition or agreement against the making of improvements without licence or consent, such covenant condition or agreement shall be deemed, notwithstanding any express provision to the contrary, to be subject to a proviso that such licence or consent is not to be unreasonably withheld; but this proviso does not preclude the right to require as a condition of such licence or consent the payment of a reasonable sum in respect of any damage to or diminution in the value of the premises or any neighbouring premises belonging to the landlord, and of any legal or other expenses properly incurred in connection with such licence or consent nor, in the case of an improvement which does not add to the letting value of the holding, does it preclude the right to require as a condition of such licence or consent, where such a requirement would be reasonable, an undertaking on the part of the tenant to reinstate the premises in the condition in which they were before the improvement was executed.

(3) In all leases whether made before or after the commencement of this Act containing a covenant condition or agreement against the alteration of the user of the demised premises, without licence or consent, such covenant condition or agreement shall, if the alteration does not involve any structural alteration of the premises, be deemed, notwithstanding any express provision to the contrary, to be subject to a proviso that no fine or sum of money in the nature of a fine, whether by way of increase of rent or otherwise, shall be payable for or in respect of such licence or consent; but this proviso does not preclude the right of the landlord to require payment of a reasonable sum in respect of any damage to or diminution in the value of the premises or any neighbouring premises belonging to him and of any legal or other expenses incurred in connection with such licence or consent.

Where a dispute as to the reasonableness of any such sum has been determined by a court of competent jurisdiction, the landlord shall be bound to grant the licence or consent on payment of the sum so determined to be reasonable.

(4) This section shall not apply to leases of agricultural holdings within the meaning of the [Agricultural Holdings Act 1986] [which are leases in relation to which that Act applies, or to farm business tenancies within the meaning of the Agricultural Tenancies Act 1995], and paragraph (b) of subsection (1), subsection (2) and subsection (3) of this section shall not apply to mining leases.

Law of Property Act 1925

144 No fine to be exacted for licence to assign

In all leases containing a covenant, condition, or agreement against assigning, underletting, or parting with the possession, or disposing of the land or property leased without licence or consent, such covenant, condition, or agreement shall, unless the lease contains an express provision to the contrary, be deemed to be subject to a proviso to the effect that no fine or sum of money in the nature of a fine shall be payable for or in respect of such licence or consent; but this proviso does not preclude the right to require the payment of a reasonable sum in respect of any legal or other expense incurred in relation to such licence or consent.

Housing Act 1985

91 Assignment in general prohibited

(1) A secure tenancy which is—
   (a) a periodic tenancy, or
   (b) a tenancy for a term certain granted on or after 5th November 1982,
   is not capable of being assigned except in the cases mentioned in subsection (3).

(2) If a secure tenancy for a term certain granted before 5th November 1982 is assigned, then, except in the cases mentioned in subsection (3), it ceases to be a secure tenancy and cannot subsequently become a secure tenancy.

(3) The exceptions are—
   (a) an assignment in accordance with section 92 (assignment by way of exchange);
   [(b)an assignment in pursuance of an order made under—

    (i) *section 24* [section 23A or 24][682] of the Matrimonial Causes Act 1973 (property adjustment orders in connection with matrimonial proceedings),

    (ii) section 17(1) of the Matrimonial and Family Proceedings Act 1984 (property adjustment orders after overseas divorce, &c), . . .

    (iii) paragraph 1 of Schedule 1 to the Children Act 1989 (orders for financial relief against parents)[, or

    (iv) Part 2 of Schedule 5, or paragraph 9(2) or (3) of Schedule 7, to the Civil Partnership Act 2004 (property adjustment orders in connection with civil partnership proceedings or after overseas dissolution of civil partnership, etc)];]

  (c) an assignment to a person who would be qualified to succeed the tenant if the tenant died immediately before the assignment.

92 Assignments by way of exchange

(1) It is a term of every secure tenancy that the tenant may, with the written consent of the landlord, assign the tenancy to another secure tenant who satisfies the condition in subsection (2) [or to an assured tenant who satisfies the conditions in subsection (2A)].

(2) The condition is that the other secure tenant has the written consent of his landlord to an assignment of his tenancy either to the first-mentioned tenant or to another secure tenant who satisfies the condition in this subsection.

[(2A) The conditions to be satisfied with respect to an assured tenant are—

  (a) that the landlord under his assured tenancy is either the Housing Corporation, . . . a [registered social landlord] or a housing trust which is a charity; and

  (b) that he intends to assign his assured tenancy to the secure tenant referred to in subsection (1) or to another secure tenant who satisfies the condition in subsection (2).]

(3) The consent required by virtue of this section shall not be withheld except on one or more of the grounds set out in

Schedule 3, and if withheld otherwise than on one of those grounds shall be treated as given.

(4) The landlord may not rely on any of the grounds set out in Schedule 3 unless he has, within 42 days of the tenant's application for the consent, served on the tenant a notice specifying the ground and giving particulars of it.

(5) Where rent lawfully due from the tenant has not been paid or an obligation of the tenancy has been broken or not performed, the consent required by virtue of this section may be given subject to a condition requiring the tenant to pay the outstanding rent, remedy the breach or perform the obligation.

(6) Except as provided by subsection (5), a consent required by virtue of this section cannot be given subject to a condition, and a condition imposed otherwise than as so provided shall be disregarded.

SCHEDULE 3

Grounds for Withholding Consent to Assignment by Way of Exchange

Ground 1

The tenant or the proposed assignee is obliged to give up possession of the dwelling-house of which he is the secure tenant in pursuance of an order of the court, or will be so obliged at a date specified in such an order.

[Ground 1

The tenant or the proposed assignee is subject to an order of the court for the possession of the dwelling-house of which he is the secure tenant.][682]

Ground 2

Proceedings have been begun for possession of the dwelling-house of which the tenant or the proposed assignee is the secure tenant on one or more of grounds 1 to 6 in Part I of Schedule 2 (grounds on which possession may be ordered despite absence of suitable alternative accommodation), or there has been served on the tenant or the proposed assignee a notice under section 83 (notice of proceedings for possession) which specifies one or more of those grounds and is still in force.

[*Ground 2A*

Either—
(a) a relevant order or suspended Ground 2 or 14 possession order is in force, or
(b) an application is pending before any court for a relevant order, a demotion order or a Ground 2 or 14 possession order to be made,

in respect of the tenant or the proposed assignee or a person who is residing with either of them.

A "relevant order" means—

an injunction under section 152 of the Housing Act 1996 (injunctions against anti-social behaviour);

an injunction to which a power of arrest is attached by virtue of section 153 of that Act (other injunctions against anti-social behaviour);

an injunction under section 153A, 153B or 153D of that Act (injunctions against anti-social behaviour on application of certain social landlords);

an anti-social behaviour order under section 1 of the Crime and Disorder Act 1998; or

an injunction to which a power of arrest is attached by virtue of section 91 of the Anti-social Behaviour Act 2003.

A "demotion order" means a demotion order under section 82A of this Act or section 6A of the Housing Act 1988.

A "Ground 2 or 14 possession order" means an order for possession under Ground 2 in Schedule 2 to this Act or Ground 14 in Schedule 2 to the Housing Act 1988.

Where the tenancy of the tenant or the proposed assignee is a joint tenancy, any reference to that person includes (where the context permits) a reference to any of the joint tenants.]

Ground 3

The accommodation afforded by the dwelling-house is substantially more extensive than is reasonably required by the proposed assignee.

Ground 4

The extent of the accommodation afforded by the dwelling-house is not reasonably suitable to the needs of the proposed assignee and his family.

Ground 5

The dwelling-house—
    (a) forms part of or is within the curtilage of a building which, or so much of it as is held by the landlord, is held mainly for purposes other than housing purposes and consists mainly of accommodation other than housing accommodation, or is situated in a cemetery, and
    (b) was let to the tenant or a predecessor in title of his in consequence of the tenant or predecessor being in the employment of—
    the landlord,
    a local authority,
    a *new town* [development] corporation,[682]
    [a housing action trust]
    . . .
    an urban development corporation, or
    the governors of an aided school.

Ground 6

The landlord is a charity and the proposed assignee's occupation of the dwelling-house would conflict with the objects of the charity.

Ground 7

The dwelling-house has features which are substantially different from those of ordinary dwelling-houses and which are designed to make it suitable for occupation by a physically disabled person who requires accommodation of the kind provided by the dwelling-house and if the assignment were made there would no longer be such a person residing in the dwelling-house.

Ground 8

The landlord is a housing association or housing trust which lets dwelling-houses only for occupation (alone or with others) by persons whose circumstances (other than merely financial circumstances) make it especially difficult for them to satisfy their need for housing and if the assignment were made there would no longer be such a person residing in the dwelling-house.

Ground 9

The dwelling-house is one of a group of dwelling-houses which it is the practice of the landlord to let for occupation by persons with special needs and a social service or special facility is provided in close proximity to the group of dwelling-houses in order to assist persons with those special needs and if the assignment were made there would no longer be a person with those special needs residing in the dwelling-house.

[Ground 10

The dwelling-house is the subject of a management agreement under which the manager is a housing association of which at least half the members are tenants of dwelling-houses subject to the agreement, at least half the tenants of the dwelling-houses are members of the association and the proposed assignee is not, and is not willing to become, a member of the association.]

Communications Act 2003

134 Restrictions in leases and licences

(1) This section applies where provision contained in a lease, licence or other agreement relating to premises has the effect of imposing on the occupier a prohibition or restriction under which his choice of—
    (a) the person from whom he obtains electronic communications services, or particular electronic communications services, or
    (b) the person through whom he arranges to be provided with electronic communications services, or particular electronic communications services,
is confined to a person with an interest in the premises, to a person selected by a person with such an interest or to persons who are one or the other.

(2) This section also applies where—
    (a) provision contained in a lease for a year or more has

the effect of imposing any other prohibition or restriction on the lessee with respect to an electronic communications matter; or

(b) provision contained in an agreement relating to premises to which a lease for a year or more applies has the effect of imposing a prohibition or restriction on the lessee with respect to such a matter.

(3) A provision falling within subsection (1) shall have effect—

    (a) as if the prohibition or restriction applied only where the lessor, licensor or other party to the agreement has not given his consent to a departure from the requirements imposed by the prohibition or restriction; and

    (b) as if the lessor, licensor or other party were required not to withhold that consent unreasonably.

(4) A provision falling within subsection (2)(a) or (b) shall have effect—

    (a) in relation to things done inside a building occupied by the lessee under the lease, or

    (b) for purposes connected with the provision to the lessee of an electronic communications service,

as if the prohibition or restriction applied only where the lessor has not given his consent in relation to the matter in question and as if the lessor were required not to withhold that consent unreasonably.

(5) Where (whether by virtue of this section or otherwise) a provision falling within subsection (1) or (2) imposes a requirement on a lessor, licensor or party to an agreement not unreasonably to withhold his consent—

    (a) in relation to an electronic communications matter, or

    (b) to the obtaining by the occupier of premises of an electronic communications service from or through a particular person,

the question whether the consent is unreasonably withheld has to be determined having regard to all the circumstances and to the principle that no person should unreasonably be denied access to an electronic communications network or to electronic communications services.

(6) OFCOM may by order provide for this section not to apply in the case of such provisions as may be described in the order.

(7) References in this section to electronic communications matters are references to—
    (a) the provision of an electronic communications network or electronic communications service;
    (b) the connection of electronic communications apparatus to a relevant electronic communications network or of any such network to another; and
    (c) the installation, maintenance, adjustment, repair, alteration or use for purposes connected with the provision of such a network or service of electronic communications apparatus.

(8) In this section—
"alteration" has the same meaning as in the electronic communications code;
"lease" includes—
    (a) a leasehold tenancy (whether in the nature of a head lease, sub-lease or under lease) and an agreement to grant such a tenancy, and
    (b) in Scotland, a sub-lease and an agreement to grant a sub-lease,
and "lessor" and "lessee" are to be construed accordingly;
"relevant electronic communications network" means—
    (a) a public electronic communications network that is specified for the purposes of this section in an order made by the Secretary of State; or
    (b) an electronic communications network that is, or is to be, connected (directly or indirectly) to such a network.

(9) This section applies to provisions contained in leases, licences or agreements granted or entered into before the commencement of this section to the extent only that provision to that effect is contained in an order made by OFCOM.

(10) This section is not to be construed as affecting the operation of paragraph 2(3) of the electronic communications code (lessees etc bound by rights granted under code by owners).

(11) The consent of the Secretary of State is required for the making by OFCOM of an order under this section.

(12) Section 403 applies to the powers of OFCOM to make orders under this section.

(10) A statutory instrument containing an order made by OFCOM under this section shall be subject to annulment in pursuance of a resolution of either House of Parliament.

Telecommunications Act 1984

Schedule 2 The Electronic Communications Code

Interpretation of code

1(2) In this code, references to the alteration of any apparatus include references to the moving, removal or replacement of the apparatus.

References

1 LPA 1925, s. 54(2).

2 *Moat v Martin* [1950] 1 KB 175.

3 *Royal Bank of Scotland v Victoria Street* [2008] All ER (D) 31.

4 LTA 1927, s. 19 or LPA 1925, s. 144 (as appropriate).

5 LTA 1927, s. 19(1).

6 LTA 1927, s. 19(2).

7 LPA 1925, s. 144.

8 LTA 1927, s. 19(2).

9 LTA 1927, s. 19(3).

10 LTA 1927, s. 19(1).

11 LTA 1927, s. 19(2).

12 LTA 1927, s. 19(3).

13 LTA 1927, s. 19(1).

14 LTA 1927, s. 19(2).

15 LTA 1927, s. 19(3).

16 LTA 1927, s. 19.

17 *Sargeant v Macepark (Whittlebury)* [2004] 4 All ER 662.

18 *Moat v Martin* [1950] 1 KB 175.

19 For a detailed discussion see The Law Commission's report, number 141.

20 CLRA 2002, Sch. 7, para. 13.

21 LTA 1988, s. 5(3).

22 *Criminal Justice Act* 1991, s. 84.

23 *Criminal Justice and Public Order Act* 1994, s. 7(3).

24 *Immigration and Asylum Act* 1999, s. 149(3).

25 *Dong Bang Minerva (UK) v Davina* [1995] 1 EGLR 41.

26 LTA 1988, s. 1(1)(b).

27 *Clinton Cards (Essex) v Sun Alliance and London Assurance Co* [2002] 3 EGLR 19.

28 *Re Smith's Lease, Smith v Richards* [1951] 1 All ER 346.

29 *Bocardo v S&M Hotels* [1980] 1 WLR 17.

30 *Crestfort v Tesco Stores* [2005] 3 EGLR 25.

31 *Level Properties v Balls Brothers* [2007] 2 EGLR 26.

32 LTA 1927, s. 19(1D) and (1E).

33 LTA 1927, s. 19(1A).

34 *Old English Inns v Brightside*, The Times, June 30, 2004.

35 See *Crestfort v Tesco Stores* [2005] 3 EGLR 25 at para. 49.

36 *Pazgate v McGrath* [1984] 2 EGLR 130.

37 See *In re Wright* [1949] Ch 729 and LPA 1925, s. 79.

38 *In re Farrow's Bank* [1921] 2 Ch 164.

39 *Metropolitan Water Board v Solomon* [1908] 2 Ch 214.

40 *Drive Yourself Hire Co (London) v Strutt* [1954] 1 QB 250.

41 *In re Nisbet and Potts' Contract* [1906] 1 Ch 386.

42 *Hill v Harris* [1965] 2 QB 601.

43 *Villiers v Oldcorn* (1903) 20 TLR 11.

44 *Walker v Arkay Caterers* [1997] EGCS 107.

45 *Investors Compensation Scheme v West Bromwich Building Society* [1998] 1 All ER 98 at p. 114f.

46 *Investors Compensation Scheme v West Bromwich Building Society* [1998] 1 All ER 98.

47 See *Mannai Investments Co v Eagle Star Life Assurance Co* [1997] AC 749.

48 *Antaios Compania Naviera SA v Salen Rederierna AB* [1985] AC 191 at p. 201.

49 (1838 edition), vol. III, p. 60.

50 *Sirius International Insurance Co (Publ) v FAI General Insurance* [2004] 1 WLR 3251.

51 *Old Grovebury Manor Farm v W Seymour Plant Sales & Hire (No. 2)* [1979] 3 All ER 504.

52 *Pinhorn v Souster* (1853) 8 Exch 763 (assignment); *Birch v Wright* (1786) 1 Term Rep 378 (subletting).

53 *Varley v Coppard* (1872) LR 7 CP 505; *Burton v Camden London Borough Council* [2000] 2 AC 399. The landlord can join in the transaction and release the outgoing tenant, leaving the tenancy vested in the remaining tenant(s). But as the landlord will be joining in, no problems of breach of covenant or the need for consent will arise.

54 *McEachen v Colton* [1902] AC 104.

55 *Pazgate v McGrath* [1984] 2 EGLR 130.

56 *Doe d. Mitchinson v Carter* (1798) 8 Term Rep 57.

57 *Slipper v Tottenham and Hampstead Junction Railway Co* (1867) 4 LR Eq 112.

58 *Marsh v Gilbert* [1980] 2 EGLR 44.

59 *Seers v Hind* (1791) 1 Ves 294.

60 *In re Riggs* [1901] 2 KB 16; *Cohen v Popular Restaurants* [1917] 1 KB 480.

61 *Crusoe d. Blencowe v Bugby* (1771) 3 Wils KB 234.

62 See, for example, *Parc Battersea v Hutchinson* [1999] 2 EGLR 33.

63 *Crago v Julian* [1992] 1 WLR 372.

64 *Brown & Root Technology v Sun Alliance and London Assurance Co* [2001] 1 Ch 733.

65 *M'Kay v M'Nally* (1879) 4 LR Ir 438.

66 *Crusoe d. Blencowe v Bugby* (1771) 3 Wils KB 234.

67 *Doe d. Holland v Worsely* (1807) 1 Camp 20.

68 *Gian Singh & Co v Nahar* [1965] 1 All ER 768.

69 *Gentle v Faulkner* [1906] 2 QB 267.

70 *Horsey Estate v Steiger* [1899] 2 QB 79.

71 *Grove v Portal* [1902] 1 Ch 727.

72 *Field v Barkworth* [1986] 1 All ER 362.

73 *Milmo v Carreras* [1946] KB 306; *Parc Battersea v Hutchinson* [1999] 2 EGLR 33.

74 *Grosvenor Estate Belgravia v Cochran* [1991] 2 EGLR 83.

75 *In re Doyle & O'Hara's Contract* [1899] 1 IR 113, although see the doubts expressed in *Marks v Warren* [1979] 1 All ER 29 at p. 31.

76 *Greenaway v Adams* (1806) 12 Ves 395.

77 *Doe d. Pitt v Laming* (1814) 4 Camp 73; *Porter v Gibbons* (1904) 48 SJ 559.

78 *Grand Junction Co v Bates* [1954] 2 QB 160.

79 *Horsey Estate v Steiger* [1989] 2 QB 79.

80 *Chatterton v Terrell* [1923] AC 578; *Yorkshire Metropolitan Properties v Co-operative Retail Services* [2001] L&TR 298.

81 *Wilson v Rosenthal* (1906) 22 TLR 233.

82 *Cook v Shoesmith* [1951] 1 KB 752.

83 *Esdaile v Lewis* [1956] 1 WLR 709.

84 *Field v Barkworth* [1986] 1 All ER 362.

85 *Corporation of Bristol v Westcott* (1879) 12 Ch D 461.

86 *Marks v Warren* [1979] 1 All ER 29.

87 *Horsey Estate v Steiger* [1899] 2 QB 79.

88 *Southern Depot Co v British Railways Board* [1990] 2 EGLR 39.

89 *Lam Kee Ying Sdn Bhd v Lam Shes Tong* [1975] AC 247.

90 *Church v Brown* (1808) 15 Ves 258.

91 *Russell v Beecham* [1924] 1 KB 525.

92 *Akici v LR Butlin* [2006] 1 WLR 201.

93 *Akici v LR Butlin* [2006] 1 WLR 201; *Tulapam Properties v De Almeida* [1981] 2 EGLR 55 should now be ignored on this point.

94 *Wallace v C Brian Barratt & Son* [1997] 2 EGLR 1.

95 *Mean Fiddler Holdings v Islington Borough Council* [2003] 2 EGLR 7.

96 *Willmott v London Road Car Company* [1910] 2 Ch 525.

97 HA 1985, s. 91(1).

98 As defined by HA 1985, s. 5.

99 HA 1985, s. 109.

100 HA 1985, s. 91(3)(a). The rules for assignments by way of exchange are set out in s. 92.

101 HA 1985, s. 91(3)(c).

102 *Governors of the Peabody Donation Fund v Higgins* [1983] 3 All ER 122.

103 HA 1985, s. 95.

104 HA 1996, ss. 134 and 143K.

105 *Keeves v Dean* [1924] 1 KB 685.
106 RA 1977, s. 3(5) and Sch. 1, Part 2, para. 13.
107 *Oak Property v Chapman* [1947] 2 All ER 1.
108 *Norman v Simpson* [1946] 1 All ER 74.
109 RA 1977, s. 137(2).
110 RA 1977, s. 98 and Sch. 15, Part 1, Case 1.
111 *Trustees of Henry Smith's Charity v Willson* [1983] QB 316.
112 HA 1980, s. 52.
113 HA 1980, s. 54.
114 HA 1988, s. 15.
115 HA 1988, s. 5(3)(e).
116 HA 1988, s. 15(2).
117 HA 1988, s. 15.
118 HA 1985, s. 91(2).
119 HA 1985, s. 95.
120 HA 1985, s. 93(2).
121 HA 1985, s. 95.
122 HA 1985, s. 93(1)(b).
123 HA 1985, s. 94(2).
124 HA 1985, s. 84 and Sch. 2, Part 1, Ground 1.
125 HA 1985, s. 93(1)(a).
126 *Crestfort v Tesco Stores* [2005] 3 EGLR 25.
127 LTA 1927, s. 19(1)(a).
128 *West v Gwynne* [1911] 2 Ch 1.
129 LTA 1927, s. 19(2).
130 See T. M. Fancourt QC, *Enforceability of Landlord and Tenant Covenants* (2nd edition), Sweet & Maxwell, 2006, 24.13.
131 *Sargeant v Macepark (Whittlebury)* [2004] 4 All ER 662.
132 LTA 1927, s. 19(1)(b).
133 Compare *Roberts v Church Commissioners for England* [1972] 1 QB 278 and *Cadogan (Earl of) Cadogan v Guinness* [1936] Ch 515, both on different wording.
134 LTA 1927, s. 19(1)(b).
135 LTA 1927, s. 19(4).
136 LTA 1927, s. 19(1E).
137 LTA 1927, s. 19(1D) and (1E).
138 *Vaux Group v Lilley* [1991] 1 EGLR 60.
139 *Mount Eden Land v Towerstone* [2003] L&TR 41.
140 *Iqbal v Thakrar* [2004] 3 EGLR 21.
141 *Bickmore v Dimmer* [1903] 1 Ch 158.
142 *Bickmore v Dimmer* [1903] 1 Ch 158.
143 *Joseph v London County Council* (1914) 111 LT 276.
144 *Heard v Stuart* (1907) 24 TLR 104.
145 *Hagee (London) v Co-operative Insurance Society* [1992] 1 EGLR 57.
146 *London County Council v Hutter* [1925] Ch 626.

147 *Hagee (London) v Co-operative Insurance Society* [1992] 1 EGLR 57.
148 *Pearlman v Keepers and Governors of Harrow School* [1979] All ER 365.
149 *Pearlman v Keepers and Governors of Harrow School* [1979] All ER 365.
150 *Bent v High Cliff Developments* [1999] All ER (D) 963.
151 *Pearl Assurance v Shaw* [1985] 1 EGLR 92.
152 *Doe d. Wetherell v Bird* (1833) 6 C&P 195.
153 *British Glass Manufacturers Confederation v University of Sheffield* [2004] 1 EGLR 41.
154 *Marsden v Edward Heyes* [1927] 2 KB 1.
155 *Mancetter Developments v Garmanston* [1986] 1 All ER 449.
156 *Hyman v Rose* [1912] AC 623.
157 HA 1985, s. 109; see s. 5 for the definition of co-operative housing association.
158 HA 1985, s. 97.
159 HA 1985, s. 97(3).
160 HA 1985, ss. 99A and 100.
161 HA 1985, s. 101.
162 HA 1980, s. 81.
163 Compare *Crestfort v Tesco Stores* [2005] 3 EGLR 25 on an alienation covenant.
164 LTA 1927, s. 19(2).
165 *FW Woolworth & Co v Lambert* [1937] Ch 37.
166 *Davies v Yadegar* [1990] 1 EGLR 71.
167 *Lambert v FW Woolworth & Co* (No. 2) [1938] 1 Ch 883.
168 *Lilley & Skinner v Crump* (1929) 73 SJ 366.
169 *Balls Brothers v Sinclair* [1931] 2 Ch 325.
170 *Tideway Investment and Property Holdings v Wellwood* [1952] 1 All ER 1142.
171 *Lambert v FW Woolworth & Co* (No. 2) [1938] 1 Ch 883.
172 Compare *West v Gwynne* [1911] 2 Ch 1 on an alienation covenant.
173 LTA 1927, s. 19(2).
174 CLRA 2002, Sch. 7, para. 1.
175 *Lambert v FW Woolworth & Co* (No. 2) [1938] 1 Ch 883.
176 *Lambert v FW Woolworth & Co* (No. 2) [1938] 1 Ch 883.
177 LTA 1927, s. 19(2).
178 *Sargeant v Macepark (Whittlebury)* [2004] 4 All ER 662.
179 *Disability Discrimination Act* 1995, s. 18A.
180 *Disability Discrimination Act* 1995, s. 27.
181 *Communications Act* 2003, s. 134(4).
182 S. 134(5).
183 S. 134(2).
184 S. 134(9).

185 S. 134(6).
186 S. 134(4).
187 *Global Telecommunications Law and Practice* (looseleaf), General Editors: Colin Long, Natasha Hobday, Sweet & Maxwell, UK-713.
188 Contained in Sch. 2 to the *Telecommunications Act* 1984 as amended by Sch. 3 to the *Communications Act* 2003.
189 *Communications Act* 2003, s. 134(3).
190 S. 134(5).
191 S. 134(9).
192 S. 134(6).
193 LTA 1927, s. 17.
194 *Brighton College v Marriott* [1925] 1 KB 312, [1926] AC 192.
195 LTA 1927, s. 3(1).
196 See *National Electric Theatres v Hudgell* [1939] Ch 553.
197 *National Electric Theatres v Hudgell* [1939] Ch 553.
198 *Owen Owen Estate v Livett* [1956] 1 Ch 1.
199 Kirk Reynolds QC and Wayne Clark, *Renewal of Business Tenancies* (3rd edition), Sweet & Maxwell, 2007, 13-48.
200 Set out in LTA 1927, s. 3.
201 *Deerfield Travel Services v Wardens and Society of the Mistery or Art of the Leathersellers of the City of London* [1982] 2 EGLR 39.
202 *Deerfield Travel Services v Wardens and Society of the Mistery or Art of the Leathersellers of the City of London* [1982] 2 EGLR 39.
203 *Norfolk Capital Group v Cadogan Estates* [2004] 1 WLR 1458.
204 LTA 1927, s. 1.
205 HA 1985, s. 610.
206 *Stack v Church Commissioners for England* [1952] 1 All ER 1352.
207 *Alliance Economic Investment Co v Berton* (1923) 92 LJKB 750.
208 *Access to Neighbouring Land Act* 1992, s. 1(4).
209 *Access to Neighbouring Land Act* 1992, s. 1(5).
210 *Doe d. Bute (Marquis) v Guest* (1846) 15 M & W 160.
211 *Edler v Auerbach* [1950] 1 KB 359.
212 *Sexual Offences Act* 1956, s. 35 and Sch. 1.
213 *Mander v Falcke* [1891] 2 Ch 554.
214 *Sunday Trading Act* 1994, s. 3.
215 *Shops Act* 1950.
216 *Montross Associated Investments SA v Moussaieff* [1990] 2 EGLR 61. On appeal, the question was left open: [1992] 1 EGLR 55.
217 *Montross Associated Investments SA v Moussaieff* [1992] 1 EGLR 55.
218 *Skillion v Keltec Industrial Research* [1992] 1 EGLR 123.
219 *Levermore v Jobey* [1956] 1 WLR 697.
220 *Joint London Holdings v Mount Cook Land* [2005] 3 EGLR 119.
221 *Joint London Holdings v Mount Cook Land* [2005] 3 EGLR 119.

222 *Bier v Danser* (1951) 157 EG 552.

223 *Brewers Company v Viewplan* [1989] 2 EGLR 133.

224 *Brewers Company v Viewplan* [1989] 2 EGLR 133.

225 *Harrison v Good* (1871) LR 11 Eq 338.

226 *Tod-Heatley v Benham* (1889) 40 Ch D 80.

227 *British Petroleum Pension Trust v Behrent* [1985] 2 EGLR 97.

228 *Upfill v Wright* [1911] 1 KB 506.

229 *Heglibiston Establishment v Heyman* (1977) 36 P&CR 351.

230 *Equality Act* 2006, s. 47 and *Equality Act (Sexual Orientation) Regulations* 2007.

231 *Devonshire (Duke of) v Brookshaw* (1899) 81 LT 83.

232 *Montross Associated Investments SA v Moussaieff* [1990] 2 EGLR 61 (affirmed [1992] 1 EGLR 55).

233 *Labone v Litherland Urban District Council* [1956] 1 WLR 522.

234 *Williams v Kiley* [2003] L&TR 303; cf. *St Marylebone Property Co v Tesco Stores* [1988] 2 EGLR 40.

235 *Doe d. Wetherell v Bird* (1834) 2 Ad & El 161; *Wheatley v Smithers* [1906] 2 KB 321, reversed on other grounds [1907] 2 KB 684.

236 *Rolls v Miller* (1884) 27 Ch D 71; *Florent v Horez* (1984) 48 P&CR 166.

237 *Florent v Horez* (1984) 48 P&CR 166.

238 *Thorn v Madden* [1925] Ch 847; *Tendler v Sproule* [1947] 1 All ER 193; *Barton v Keeble* [1928] Ch 517; *Bagettes v GP Estates* [1956] Ch 290.

239 *Tubbs v Esser* (1909) 26 TLR 145.

240 *Dobbs v Linford* [1953] 1 QB 48; *Crest Nicholson Residential (South) v McAllister* [2003] 1 All ER 46 (reversed in the Court of Appeal on other grounds).

241 *Martin v David Wilson Homes* [2004] 3 EGLR 77.

242 *Day v Waldron* (1919) 88 LJKB 937; *Belton v Nicholl* [1941] IR 230.

243 *Berton v London and Counties House Property Co,* unreported, 17 November 1920 (cited in *Berton v Alliance Economic Investment Co* [1922] 1 KB 742).

244 *Thorn v Madden* [1925] Ch 847; *Tendler v Sproule* [1947] 1 All ER 193; *Segal Securities v Thoseby* [1963] 1 QB 887.

245 *Falgor Commercial v Alsabahia* [1986] 1 EGLR 41.

246 *Stoke Lands v Sears* (1948) 151 EG 196.

247 *Caradon District Council v Paton* [2000] 3 EGLR 57.

248 *Roberts v Howlett* [2002] 1 P&CR 19.

249 *Reeves v Cattell* (1876) 24 WR 485.

250 Contrast *Downie v Turner* [1951] 2 KB 112 with *Dobbs v Linford* [1953] 1 QB 48.

251 *St Marylebone Property Co v Tesco Stores* [1988] 2 EGLR 40.

252 *Calabar (Woolwich) v Tesco Stores* [1978] 1 EGLR 113.

253 *Basildon Development Corporation v Mactro* [1986] 1 EGLR 137.

254 *Bridgegrove v Smith* [1997] 2 EGLR 40.

255 *Atual v Courts Garages* [1989] 1 EGLR 63.

256 *Pearl Assurance v Shaw* [1985] 1 EGLR 92.

257 Compare *Crestfort v Tesco Stores* [2005] 3 EGLR 25 on an alienation covenant.

258 LTA 1927, s. 19(3).

259 Letitia Crabb and Jonathan Seitler QC, *Leases: Covenants and Consents* (2nd edition), Sweet & Maxwell, 2008, 9.48; Law Com No. 141 para. 3.26.

260 LTA 1927, s. 19(3).

261 CLRA 2002, Sch. 7, para. 1.

262 Compare *Lambert v FW Woolworth & Co* (No. 2) [1938] 1 Ch 883 on alterations.

263 LTA 1927, s. 19(3).

264 LTA 1927, s. 19(3).

265 *Barclays Bank v Daejan Investments (Grove Hall)* [1995] 1 EGLR 68.

266 *Guardian Assurance Co v Gants Hill Holdings* [1983] 2 EGLR 36.

267 See, for example, *Mahon v Sims* [2005] 3 EGLR 67 and *Town Quay Developments v Eastleigh Borough Council* [2008] EWHC 1922 (Ch).

268 Compare *West v Gwynne* [1911] 2 Ch 1 on an alienation covenant.

269 *Leasehold Reform, Housing and Urban Development Act* 1993, s. 89.

270 LPA 1925, s. 84.

271 *Westminster City Council v Westminster (Duke)* [1991] 4 All ER 136.

272 LPA 1925, s. 84(12).

273 *Cadogan (Earl of) v Guinness* [1936] Ch 515.

274 LPA 1925, s. 84(12).

275 *Cadogan (Earl of) v Guinness* [1936] Ch 515.

276 LPA 1925, s. 84(12).

277 In ss. 84(7) and 87(11).

278 *Ridley v Taylor* [1965] 2 All ER 51.

279 LPA 1925, s. 84(1).

280 LPA 1925, s. 84(1).

281 LPA 1925, s. 84(1C).

282 LPA 1925, s. 84(9).

283 *Iveagh v Harris* [1929] 2 Ch 142.

284 *Driscoll v Church Commissioners for England* [1957] 1 QB 330.

285 *Creery v Summersell and Flowerdew & Co* [1949] Ch 751.

286 Compare, for example, *Lay v Ackerman* [2005] 1 EGLR 139 and *Havant International v Lionsgate (H) Investment* [2000] 2 L&TR 297 in the context of a tenant's break notice.

287 See, for example, *Stait v Fenner* [1912] 2 Ch 504.

288 *Land Registration Act* 2002, s. 27(1); *Land Registration Act* 1925, s. 22(1).

289 LTA 1987, s. 47.

290 CLRA 2002, Part 2.

291 CLRA 2002, s. 98.

292 CLRA 2002, s. 111.

293 CLRA 2002, ss. 76 and 77.

294 CLRA 2002, s. 98(2).

295 See in particular CLRA 2002, s. 98(6).

296 See CLRA 2002, s. 75.

297 *NCR v Riverland Portfolio* (No. 2) [2005] 2 EGLR 42.

298 *Norwich Union Life Insurance Society v Shopmoor* [1998] 3 All ER 32.

299 *Royal Bank of Scotland v Victoria Street* [2008] All ER (D) 31.

300 *Allied Dunbar Assurance v Homebase* [2002] 2 EGLR 23.

301 *Norwich Union Linked Life Assurance v Mercantile Credit Co* [2003] EWHC 3064 (Ch).

302 *Footwear Corporation v Amplight Properties* [1999] 1 WLR 551.

303 See *Venetian Glass Gallery v Next Properties* [1989] 2 EGLR 42 and *Old English Inns v Brightside*, The Times, June 30, 2004.

304 *Ponderosa International Developments v Pengap Securities (Bristol)* [1986] 1 EGLR 66.

305 *Ponderosa International Development v Pengap Securities (Bristol)* [1986] 1 EGLR 66; *British Bakeries (Midlands) v Michael Testler & Co* [1986] 1 EGLR 64.

306 *Ponderosa International Development v Pengap Securities (Bristol)* [1986] 1 EGLR 66.

307 *British Bakeries (Midlands) v Michael Testler & Co* [1986] 1 EGLR 64.

308 *Ponderosa International Development v Pengap Securities (Bristol)* [1986] 1 EGLR 66.

309 *British Bakeries (Midlands) v Michael Testler & Co* [1986] 1 EGLR 64.

310 *British Bakeries (Midlands) v Michael Testler & Co* [1986] 1 EGLR 64.

311 *Fullers Theatre & Vaudeville Co v Rofe* [1923] AC 435.

312 *Allied Dunbar Assurance v Homebase* [2002] 2 EGLR 23.

313 *Kened v Connie Investments* [1997] 1 EGLR 21 and see *Dong Bang Minerva (UK) v Davina* [1995] 1 EGLR 41.

314 *Iqbal v Thakrar* [2004] 3 EGLR 21.

315 LTA 1927, s. 19(3).

316 *Dong Bang Minerva (UK) v Davina* [1996] 2 EGLR 31.

317 *Aubergine Enterprises v Lakewood International* [2002] 1 WLR 2149.

318 *International Drilling Fluids v Louisville Investments (Uxbridge)* [1986] Ch 513 (assignment); *Iqbal v Thakrar* [2004] 3 EGLR 21 (alterations); *Tollbench v Plymouth City Council* [1988] 1 EGLR 79 (change of use).

319 LTA 1988, s. 5(2).

320 LPA 1925, s. 196(3) and (4).

321 LPA 1925, s. 196(4).

322 LTA 1987, s. 49.

323 LTA 1988, s. 5(2).

324 LTA 1927, s. 23(1).

325 LTA 1988, s. 5(2).

326 See *Galinski v McHugh* [1989] 1 EGLR 109.

327 *Tanham v Nicholson* (1871) LR 5 HL 561.

328 *Townsend Carriers v Pfizer* (1977) 33 P&CR 361.

329 *Norwich Union Linked Life Assurance v Mercantile Credit Co* [2003] EWHC 3064 (Ch).

330 *Isow's Restaurants v Greenhaven (Piccadilly) Properties* (1969) 213 EG 505.

331 LTA 1988, s. 2(1).

332 *Mackusick v Carmichael* [1917] 2 KB 581.

333 LTA 1988, s. 2(1).

334 LTA 1988, s. 2(1).

335 We consider that the statement in *The Landlord and Tenant Factbook* (looseleaf), Editors: Matthew Marsh and Zia Bhaloo, Sweet & Maxwell, para. 4B-10.4, that the duty applies to the landlord's professional advisers, is incorrect as they are not persons 'who may consent to the transaction'.

336 LTA 1988, s. 2(2).

337 LTA 1988, s. 3.

338 LTA 1988, s. 3(2)(b).

339 LTA 1988, s. 1(3).

340 LTA 1988, s. 1(3)(b).

341 LTA 1988, s. 1(4) and *Young v Ashley Gardens Properties* [1903] 2 Ch 112.

342 *Go West v Spigarolo* [2003] QB 1140.

343 *Footwear Corporation v Amplight Properties* [1999] 1 WLR 551.

344 *Ashworth Frazer v Gloucester City Council* (No. 2) [2002] 1 All ER 337.

345 *CIN Properties v Gill* [1993] 3 EGLR 97.

346 *BRS Northern v Templeheights* [1998] 2 EGLR 182.

347 *Go West v Spigarolo* [2003] QB 1140.

348 *Go West v Spigarolo* [2003] QB 1140.

349 *NCR v Riverland Portfolio* (No. 2) [2005] 2 EGLR 42.

350 *Go West v Spigarolo* [2003] QB 1140.

351 *NCR v Riverland Portfolio* (No. 2) [2005] 2 EGLR 42.

352 *Norwich Union Life Insurance Society v Shopmoor* [1998] 3 All ER 32.

353 *Go West v Spigarolo* [2003] QB 1140.

354 *Mount Eden Land v Folia* [2003] EWHC 1815 (Ch).

355 *Blockbuster Entertainment v Barnsdale Properties* [2004] L&TR 239.

356 *NCR v Riverland Portfolio* (No. 2) [2005] 2 EGLR 42.

357 *Go West v Spigarolo* [2003] QB 1140.

358 *Go West v Spigarolo* [2003] QB 1140.

359 S. 5(3).

360 HA 1985, s. 92(3) and (4).

361 HA 1985, s. 94(6).

362 *City Hotels Group v Total Property Investments* [1985] 1 EGLR 253.

363 *Lewis & Allenby (1909) v Pegge* [1914] 1 Ch 782.

364 *City Hotels Group v Total Property Investments* [1985] 1 EGLR 253.

365 *Young v Ashley Gardens Properties* [1903] 2 Ch 112.

366 *Frederick Berry v Royal Bank of Scotland* [1949] 1 KB 619.

367 *Sonnenthal v Newton* (1965) 109 SJ 333; *Welch v Birrane* (1974) 29 P&CR 102.

368 *Bromley Park Garden Estates v Moss* [1982] 1 WLR 1019.

369 *Treloar v Bigge* (1874) LR 9 Ex 151.

370 *Treloar v Bigge* (1874) LR 9 Ex 151.

371 *Ideal Film Renting Co v Neilsen* [1921] 1 Ch 575.

372 *Disability Discrimination Act* 1995, s. 18A.

373 *Disability Discrimination Act* 1995, s. 27.

374 *Disability Discrimination (Employment Field) (Leasehold Premises) Regulations* 2004, reg. 4.

375 *Disability Discrimination (Providers of Services) (Adjustment of Premises) Regulations* 2001, reg. 5.

376 *Disability Discrimination (Employment Field) (Leasehold Premises) Regulations* 2004, reg. 5 and *Disability Discrimination (Providers of Services) (Adjustment of Premises) Regulations* 2001, reg. 6.

377 HA 1980, s. 82(3) and HA 1985, s. 98(4).

378 CLRA 2002, s. 98.

379 CLRA 2002, s. 98(4).

380 CLRA 2002, s. 99(1).

381 CLRA 2002, s. 99(4).

382 CLRA 2002, s. 99(2).

383 *Norwich Union Life Insurance Society v Shopmoor* [1998] 3 All ER 32.

384 *Aubergine Enterprises v Lakewood International* [2002] 1 WLR 2149.

385 *Dong Bang Minerva (UK) v Davina* [1996] 2 EGLR 31.

386 *Norwich Union Linked Life Assurance v Mercantile Credit Co* [2003] EWHC 3064 (Ch), *Allied Dunbar Assurance v Homebase* [2002] 2 EGLR 23.

387 *Allied Dunbar Assurance v Homebase* [2002] 2 EGLR 23.

388 *Kened v Connie Investments* [1997] 1 EGLR 21.

389 *Kened v Connie Investments* [1997] 1 EGLR 21.

390 *British Bakeries (Midlands) v Michael Testler & Co* [1986] 1 EGLR 64.

391 *Mount Eden Land v Towerstone* [2003] L&TR 41.

392 *Old English Inns v Brightside,* The Times, June 30, 2004.

393 *Footwear Corporation v Amplight Properties* [1999] 1 WLR 551.

394 *Design Progression v Thurloe Properties* [2005] 1 WLR 1.

395 *Tomlin v Tichler* (1952) 159 EG 602.

396 *Tollbench v Plymouth City Council* [1988] 1 EGLR 79; *Ashworth Frazer v Gloucester City Council* (No. 2) [2002] 1 All ER 337.

397 *Iqbal v Thakrar* [2004] 3 EGLR 21.

398 See, for example, *Next v National Farmers Union Mutual Insurance Co* [1997] EGCS 181.

399 *Allied Dunbar Assurance v Homebase* [2002] 2 EGLR 23, on the first of the three examples.

400 *NCR v Riverland Portfolio* [2004] L&TR 537.

401 *Re Smith's Lease, Smith v Richards* [1951] 1 All ER 346.

402 LTA 1927, s. 19(1A).

403 *Berenyi v Watford Borough Council* [1980] 2 EGLR 38.

404 [1986] Ch 513.

405 (1996) 74 P&CR 306.

406 *NCR v Riverland Portfolio* (No. 2) [2005] 2 EGLR 42.

407 See *Iqbal v Thakrar* [2004] 3 EGLR 21 (alterations) and *Tollbench v Plymouth City Council* [1988] 1 EGLR 79 (change of use).

408 *Royal Bank of Scotland v Victoria Street* [2008] All ER (D) 31.

409 *Sargeant v Macepark (Whittlebury)* [2004] 4 All ER 662.

410 *Sargeant v Macepark (Whittlebury)* [2004] 4 All ER 662.

411 *Tollbench v Plymouth City Council* [1988] 1 EGLR 79.

412 *BRS Northern v Templeheights* [1998] 2 EGLR 182.

413 *Lehmann v McArthur* (1867) 3 Eq 746.

414 *Bromley Park Garden Estates v Moss* [1982] 1 WLR 1019.

415 *Bates v Donaldson* [1896] 2 QB 241.

416 *Olympia & York Canary Wharf v Oil Property Investments* [1994] 2 EGLR 48.

417 *Rayburn v Wolf* [1985] 2 EGLR 235.

418 *Crown Estates Commissioners v Signet Group* [1996] 2 EGLR 200.

419 *Yorkshire Metropolitan Properties v Co-operative Retail Services* [2001] L&TR 298 and *Mount Eden Land v Folia* [2003] EWHC 1815 (Ch).

420 *Straudley Investments v Mount Eden Land* (No. 2), unreported, 8 December 1997.

421 *Vienit v W Williams & Son (Bread Street)* [1958] 3 All ER 621.

422 *NCR v Riverland Portfolio* (No. 2) at first instance, quoted on appeal at [2005] 2 EGLR 42.

423 *Norwich Union Life Insurance Society v Shopmoor* [1998] 3 All ER 32.

424 *Norwich Union Life Insurance Society v Shopmoor* [1998] 3 All ER 32.

425 *Blockbuster Entertainment v Leakcliff Properties* [1997] 1 EGLR 28.

426 *Luminar Leisure v Apostle* [2001] 3 EGLR 23.

427 *Premier Rinks v Amalgamated Cinematograph Theatres* (1912) 56 SJ 536; *Jaison Property Development Co v Roux Restaurants* (1996) 74 P&CR 357.

428 *Sargeant v Macepark (Whittlebury)* [2004] 4 All ER 662.

429 *Straudley Investments v Mount Eden Land* (1996) 74 P&CR 306.

430 *City Hotels Group v Total Property Investments* [1985] 1 EGLR 253.

431 *Rose v Stavrou* [2000] L&TR 133.

432 See, for example, *British Bakeries (Midlands) v Michael Testler & Co* [1986] 1 EGLR 64.

433 *Royal Bank of Scotland v Victoria Street* [2008] EWHC 3052 (Ch).

434 *Evans v Dawkins* [1968] CLY 2162.

435 *Ideal Film Renting Co v Nielsen* [1921] 1 Ch 575.

436 *Landlord Protect v Dolman* [2007] 2 EGLR 21.

437 *Old English Inns v Brightside*, The Times, June 30, 2004.

438 *Design Progression v Thurloe Properties* [2005] 1 WLR 1.

439 Contrast *British Bakeries (Midlands) v Michael Testler & Co* [1986] 1 EGLR 64 with *Footwear Corporation v Amplight Properties* [1999] 1 WLR 551.

440 *Ponderosa International Development v Pengap Securities (Bristol)* [1986] 1 EGLR 66.

441 *Ponderosa International Development v Pengap Securities (Bristol)* [1986] 1 EGLR 66.

442 *Ponderosa International Development v Pengap Securities (Bristol)* [1986] 1 EGLR 66.

443 *In re Greater London Properties' Lease* [1959] 1 WLR 503.

444 *British Bakeries (Midlands) v Michael Testler & Co* [1986] 1 EGLR 64.

445 *British Bakeries (Midlands) v Michael Testler & Co* [1986] 1 EGLR 64.

446 *Ponderosa International Development v Pengap Securities (Bristol)* [1986] 1 EGLR 66.

447 *Design Progression v Thurloe Properties* [2005] 1 WLR 1.

448 *Design Progression v Thurloe Properties* [2005] 1 WLR 1.

449 *Venetian Glass Gallery v Next Properties* [1989] 2 EGLR 42.

450 *Warren v Marketing Exchange for Africa* [1988] 2 EGLR 247, citing *Geland Manufacturing Co v Levy Estates Co* (1962) CLY 1700.

451 *Warren v Marketing Exchange for Africa* [1988] 2 EGLR 247.

452 *Ponderosa International Development v Pengap Securities (Bristol)* [1986] 1 EGLR 66.

453 *Ponderosa International Development v Pengap Securities (Bristol)* [1986] 1 EGLR 66.

454 *Design Progression v Thurloe Properties* [2005] 1 WLR 1.

455 *Royal Bank of Scotland v Victoria Street* [2008] EWHC 3052 (Ch).

456 *Ponderosa International Development v Pengap Securities (Bristol)* [1986] 1 EGLR 66.

457 *Footwear Corporation v Amplight Properties* [1999] 1 WLR 551.

458 *NCR v Riverland Portfolio* (No. 2) [2005] 2 EGLR 42.

459 *Norwich Union Life Insurance Society v Shopmoor* [1998] 3 All ER 32.

460 *Footwear Corporation v Amplight Properties* [1999] 1 WLR 551.

461 *Footwear Corporation v Amplight Properties* [1999] 1 WLR 551.

462 *Footwear Corporation v Amplight Properties* [1999] 1 WLR 551.

463 Contrast *Footwear Corporation v Amplight Properties* [1999] 1 WLR 551 with *British Bakeries (Midlands) v Michael Testler & Co* [1986] 1 EGLR 64.

464 *Design Progression v Thurloe Properties* [2005] 1 WLR 1.

465 *Kened v Connie Investments* [1997] 1 EGLR 21.

466 *Lee v K Carter* [1949] 1 KB 85.

467 *Re Cooper's Lease, Cowan v Beaumont Property Trusts* (1968) 19 P&CR 541.

468 *Norfolk Capital Group v Kitway* [1977] 1 EGLR 26.

469 *Race Relations Act* 1976, s. 57; *Sex Discrimination Act* 1975, s. 66; *Equality Act* 2006, s. 81 and *Equality Act (Sexual Orientation) Regulations* 2007, reg. 20; *Equality Act* 2006, s. 66.

470 *Race Relations Act* 1976, ss. 1 and 24.

471 *Sex Discrimination Act* 1975, ss. 1, 2, 2A, 3B and 31.

472 *Equality Act* 2006, s. 81 and *Equality Act (Sexual Orientation) Regulations* 2007, regs. 3 and 5.

473 *Equality Act* 2006, ss. 45 and 47.

474 *Race Relations Act* 1976, s. 24(4) and (5); *Sex Discrimination Act* 1975, s. 31(3) and (4).

475 *Equality Act* 2006, s. 47(3); *Equality Act* 2006, s. 81 and *Equality Act (Sexual Orientation) Regulations* 2007, reg. 5(3).

476 *Disability Discrimination Act* 1995, s. 22(4).

477 *Disability Discrimination Act* 1995, s. 25.

478 *Disability Discrimination Act* 1995, s. 22(5) and (6).

479 *Disability Discrimination Act* 1995, s. 24(6).

480 *Kened v Connie Investments* [1997] 1 EGLR 21.

481 *Parker v Boggan* [1947] KB 346.

482 *Norwich Union Life Insurance Society v Shopmoor* [1998] 3 All ER 32.

483 *In re Town Investments Underlease* [1954] Ch 301.

484 *Balfour v Kensington Gardens Mansions* (1932) 49 TLR 29.

485 *Kened v Connie Investments* [1997] 1 EGLR 21.

486 *NCR v Riverland Portfolio* (No. 2) [2005] 2 EGLR 42.

487 *Ashworth Frazer v Gloucester City Council* (No. 2) [2002] 1 All ER 377.

488 *Bates v Donaldson* [1896] 2 QB 241.

489 *International Drilling Fluids v Louisville Investments (Uxbridge)* [1986] Ch 513.

490 See, for example, *In re Spark's Lease* [1905] 1 Ch 456.

491 *Footwear Corporation v Amplight Properties* [1999] 1 WLR 551.

492 *Coopers & Lybrand v Schwartz (William) Construction Co* (1980) 116 DLR (3d) 450.

493 *Whiteminster Estates v Hodges Menswear* (1974) 232 EG 715.

494 *Premier Confectionery (London) Company v London Commercial Sale Rooms* [1933] Ch 904.

495 *International Drilling Fluids v Louisville Investments (Uxbridge)* [1986] Ch 513.

496 See *Sargeant v Macepark (Whittlebury)* [2004] 4 All ER 662.

497 *Berenyi v Watford Borough Council* [1980] 2 EGLR 38.

498 *Houlder Brothers and Company v Gibbs* [1925] Ch 575.

499 *NCR v Riverland Portfolio* (No. 2) [2005] 2 EGLR 42.

500 *NCR v Riverland Portfolio* (No. 2) [2005] 2 EGLR 42.

501 *International Drilling Fluids v Louisville Investments (Uxbridge)* [1986] Ch 513.

502 *International Drilling Fluids v Louisville Investments (Uxbridge)* [1986] Ch 513.

503 *Ponderosa International Development v Pengap Securities (Bristol)* [1986] 1 EGLR 66.

504 *International Drilling Fluids v Louisville Investments (Uxbridge)* [1986] Ch 513.

505 *Footwear Corporation v Amplight Properties* [1999] 1 WLR 551.

506 *Farr v Ginnings* (1928) 44 TLR 429.

507 *Orlando Investments v Grosvenor Estate Belgravia* [1989] 2 EGLR 74; *Crestfort v Tesco Stores* [2005] 3 EGLR 25.

508 Compare *Cosh v Fraser* (1964) 108 SJ 116 and *Goldstein v Sanders* [1915] 1 Ch 549.

509 *Crestfort v Tesco Stores* [2005] 3 EGLR 25.

510 *Straudley Investments v Mount Eden Land* (No. 2), unreported, 8 December 1997.

511 *Mount Eden Land v Folia* [2003] EWHC 1815 (Ch).

512 *Design Progression v Thurloe Properties* [2005] 1 WLR 1.

513 *Crown Estates Commissioners v Signet Group* [1996] 2 EGLR 200.

514 *Moss Bros Group v CSC Properties* [1999] EGCS 47.

515 *Oriel Property Trust v Kidd* (1949) 154 EG 500.

516 *Evans v Levy* [1910] 1 Ch 452. Compare *Landlord Protect v St Anselm* [2008] EWHC 1582 (Ch) on guarantees.

517 LT(C)A 1995, s. 25.

518 *Storehouse Properties v Ocobase*, The Times, April 3, 1998 (note that the report does not set out the facts accurately).

519 *Landlord Protect v Dolman* [2007] 2 EGLR 21.

520 *Landlord Protect v St Anselm* [2008] EWHC 1582 (Ch).

521 See *Landlord Protect v St Anselm* [2008] EWHC 1582 (Ch), where this was not challenged.

522 *Landlord Protect v St Anselm* [2008] EWHC 1582 (Ch).

523 *Landlord Protect v St Anselm* [2008] EWHC 1582 (Ch).

524 *Landlord Protect v Dolman* [2007] 2 EGLR 21.

525 LT(C)A 1995, s. 11.

526 Set out in LT(C)A 1995, s. 16.

527 LT(C)A 1995, s. 16(6).

528 *Wallis Fashion Group v CGU Life Assurance* (2001) 81 P&CR 28.

529 See *Allied Dunbar Assurance v Homebase* [2002] 2 EGLR 23.

530 *In re Spark's Lease* [1905] 1 Ch 456.

531 *Lambert v FW Woolworth & Co* (No. 2) [1938] 1 Ch 883.

532 *Iqbal v Thakrar* [2004] 3 EGLR 21.

533 *Lambert v FW Woolworth & Co* (No. 2) [1938] 1 Ch 883.

534 *Lambert v FW Woolworth & Co* (No. 2) [1938] 1 Ch 883.

535 *Mosley v Cooper* [1990] 1 EGLR 124.

536 *Iqbal v Thakrar* [2004] 3 EGLR 21.

537 *Iqbal v Thakrar* [2004] 3 EGLR 21.

538 *Crown Estates Commissioners v Signet Group* [1996] 2 EGLR 200.

539 LTA 1927, s. 19(2).

540 *Lambert v FW Woolworth & Co* (No. 2) [1938] 1 Ch 883.

541 LTA 1927, s. 19(2).

542 *Sargeant v Macepark (Whittlebury)* [2004] 4 All ER 662.

543 See *Footwear Corporation v Amplight Properties* [1999] 1 WLR 551.

544 *Tollbench v Plymouth City Council* [1988] 1 EGLR 79.

545 *Sportoffer v Erewash Borough Council* [1999] 3 EGLR 136.

546 *Sportoffer v Erewash Borough Council* [1999] 3 EGLR 136.

547 *Warren v Marketing Exchange for Africa* [1988] 2 EGLR 247.

548 *Crown Estates Commissioners v Signet Group* [1996] 2 EGLR 200.

549 *Crown Estates Commissioners v Signet Group* [1996] 2 EGLR 200.

550 *OHS v Green Property Co* [1986] IR 39.

551 Compare *Berenyi v Watford Borough Council* [1980] 2 EGLR 38.

552 *Berenyi v Watford Borough Council* [1980] 2 EGLR 38.

553 *Anglia Building Society v Sheffield City Council* [1983] 1 EGLR 57.

554 LTA 1927, s. 19(3).

555 Compare comments in *Footwear Corporation v Amplight Properties* [1999] 1 WLR 551.

556 Sch. 3 to the HA 1985.

557 HA 1985, s. 92.

558 See HA 1985, Part X.

559 HA 1985, s. 94.

560 HA 1985, s. 98(2); HA 1980, s. 82(1).

561 *Disability Discrimination (Employment Field) (Leasehold Premises) Regulations* 2004, regs. 5, 6 and 7.

562 *Disability Discrimination (Providers of Services) (Adjustment of Premises) Regulations* 2001, reg. 6.

563 *Goldman v Abbott* [1989] 2 EGLR 78.

564 *Dong Bang Minerva (UK) v Davina* [1996] 2 EGLR 31.

565 *Aubergine Enterprises v Lakewood International* [2002] 1 WLR 2149.

566 *Goldman v Abbott* [1989] 2 EGLR 78.

567 See, for example, *Design Progression v Thurloe Properties* [2005] 1 WLR 1.

568 CLRA 2002, Sch. 11.

569 LPA 1925, s. 205(1).

570 *In re Cosh's Contract* [1897] 1 Ch 9 at p. 14.

571 *Gardner & Co v Cone* [1928] 1 Ch 955.

572 *Comber v Fleet Electrics* [1955] 2 All ER 161.

573 *Hughes v Waite* [1957] 1 WLR 713 (albeit not in the context of landlords' consents).

574 *Jenkins v Price* [1907] 2 Ch 229.

575 *Barclays Bank v Daejan Investments (Grove Hall)* [1995] 1 EGLR 68.

576 *Barclays Bank v Daejan Investments (Grove Hall)* [1995] 1 EGLR 68.

577 *In re Cosh's Contract* [1897] 1 Ch 9.

578 *Roe d. Gregson v Harrison* (1788) 2 Term Rep 425.

579 *Rutter v Michael John* (1967) 201 EG 299.

580 *Next v National Farmers Union Mutual Insurance Co* [1997] EGCS 181.

581 *Venetian Glass Gallery v Next Properties* [1989] 2 EGLR 42.

582 *Bader Properties v Linley Property Investments* (1967) 19 P&CR 620.

583 *Rose v Stavrou* [2000] L&TR 133.

584 *Prudential Assurance Co v Mount Eden Land* [1997] 1 EGLR 37.

585 *Next v National Farmers Union Mutual Insurance Co* [1997] EGCS 181.

586 *Venetian Glass Gallery v Next Properties* [1989] 2 EGLR 42.

587 *Willmott v Barber* (1880) 15 Ch D 96.

588 *Richardson v Evans* (1818) 3 Madd 218.

589 *Millard v Humphreys* (1918) 62 SJ 505.

590 *Mitten v Fagg* [1978] 2 EGLR 40.

591 *Sanctuary Housing Association v Baker* [1998] 1 EGLR 42.

592 LPA 1925, s. 143.

593 See *Allied Dunbar Assurance v Homebase* [2002] 2 EGLR 23.

594 *Storehouse Properties v Ocobase*, The Times, April 3, 1998.

595 *William Hill (Southern) v Cabras* [1985] 2 EGLR 62.

596 *Westminster City Council v HSBC Bank* [2003] 1 EGLR 62.

597 See Nicholas Dowding QC and Kirk Reynolds QC, *Dilapidations: the modern law and practice* (3rd edition), Sweet & Maxwell, 2004, 17-04.

598 *Scottish Mutual Assurance Society v British Telecommunications*, unreported, 18 March 1994.

599 *Deerfield Travel Services v Wardens and Society of the Mistery or Art of the Leathersellers of the City of London* [1982] 2 EGLR 39.

600 *Rose v Stavrou* [2000] L&TR 133.

601 *System Floors v Ruralpride* [1995] 1 EGLR 48.

602 *Edler v Auerbach* [1950] 1 KB 359.

603 *Crestfort v Tesco Stores* [2005] 3 EGLR 25.

604 *Creery v Summersell and Flowerdew & Co* [1949] Ch 751.

605 LPA 1925, s. 56; *Drive Yourself Hire Co (London) v Strutt* [1954] 1 QB 250.

606 *Co-operative Insurance Society v Argyll Stores (Holdings)* [1998] AC 1.

607 *Hemingway v Dunraven* [1995] 1 EGLR 61.

608 *Crestfort v Tesco Stores* [2005] 3 EGLR 25.

609 LPA 1925, s. 84(9).

610 LPA 1925, s. 56; *Drive Yourself Hire Co (London) v Strutt* [1954] 1 QB 250.

611 *Fuller v Judy Properties* [1992] 1 EGLR 75.

612 *Central Estates (Belgravia) v Woolgar* (No. 2) [1972] 1 WLR 1048.

613 *Expert Clothing Service and Sales v Hillgate House* [1986] Ch 340 per Slade LJ.

614 *David Blackstone v Burnetts (West End)* [1973] 1 WLR 1487.

615 *Trustees of Henry Smith's Charity v Willson* [1983] QB 316.

616 *Cooper v Henderson* [1982] 2 EGLR 42.

617 *Downie v Turner* [1951] 2 KB 112.

618 *Central Estates (Belgravia) v Woolgar* (No. 2) [1972] 1 WLR 1048.

619 *Central Estates (Belgravia) v Woolgar* (No. 2) [1972] 1 WLR 1048.

620 *John Lewis Properties v Viscount Chelsea* [1993] 2 EGLR 77.

621 Compare *Milverton Group v Warner World* [1995] 2 EGLR 28.

622 *Yorkshire Metropolitan Properties v Co-operative Retail Services* [2001] L&TR 298; *Mount Eden Land v Folia* [2003] EWHC 1815 (Ch).

623 *Bader Properties v Linley Property Investments* (1967) 19 P&CR 620.

624 *Calabar Properties v Seagull Autos* [1969] 1 Ch 451.

625 *GS Fashions v B&Q* [1995] 1 WLR 1088.

626 *Criminal Law Act* 1977, s. 6.

627 *Protection From Eviction Act* 1977, s. 1.

628 *Patel v Pirabakaran* [2006] 4 All ER 506.

629 *Barrow v Isaacs & Son* [1891] 1 QB 417; *Eastern Telegraph Co v Dent* [1899] 1 QB 835.

630 See, for example, *Southern Depot Co v British Railways Board* [1990] 2 EGLR 39.

631 CLRA 2002, s. 168.

632 RA 1977, Sch. 15, Part 1.

633 RA 1977, s. 98.

634 *Regional Properties Co v Frankenschwerth and Chapman* [1951] 1 All ER 178.

635 *Finkle v Strzeleczyk* [1961] 3 All ER 409.

636 RA 1977, s. 137(2).

637 *Leith Properties v Springer* [1982] 3 All ER 731 ('Tenant' refers to the last immediate tenant of the landlord 'excluding an assignee of the tenancy or a subtenant who has not yet been accepted by the landlord as a tenant').

638 *City Hotels Group v Total Property Investments* [1985] 1 EGLR 253.

639 *Frederick Berry v Royal Bank of Scotland* [1949] 1 KB 619.

640 *Treloar v Bigge* (1874) LR 9 Ex 151.

641 *Young v Ashley Gardens Properties* [1903] 2 Ch 112.

642 *West v Gwynne* [1911] 2 Ch 1.

643 *Treloar v Bigge* (1874) LR 9 Ex 151; *Sear v House Property and Investments Society* (1880) 16 Ch D 387.

644 See *Ideal Film Renting Co v Neilsen* [1921] 1 Ch 575 on assignment for an example.

645 *Blockbuster Entertainment v Barnsdale Properties* [2004] L&TR 239.

646 *Design Progression v Thurloe Properties* [2005] 1 WLR 1.

647 *Clinton Cards (Essex) v Sun Alliance and London Assurance Co* [2002] 3 EGLR 19.

648 *Blockbuster Entertainment v Barnsdale Properties* [2004] L&TR 239.

649 *Limitation Act* 1980, s. 9.

650 *Comber v Fleet Electrics* [1955] 2 All ER 161.

651 *Andrew v Bridgman* [1908] 1 KB 596.

652 RA 1977, Sch. 1, Part 2, para. 14.

653 LT(C)A 1995, s. 1.

654 LT(C)A 1995, s. 1(5).

655 *Friends Provident Life Office v British Railways Board* [1996] 1 All ER 336.

656 HA 1988, s. 34.

657 *Law of Property (Miscellaneous Provisions) Act* 1989, s. 1; *Companies Act* 1985, s. 35AA (to be replaced, in the same terms, by *Companies Act* 2006, s. 46).

658 *West v Gwynne* [1911] 2 Ch 1, on the predecessor section.

659 LTA 1927, s. 25.

660 *Leasehold Reform Act* 1967, s. 30.

661 LTA 1927, s. 19(4).

662 LTA 1927, s. 19(4).

663 *Criminal Justice Act* 1991, s. 84.

664 *Criminal Justice and Public Order Act* 1994, s. 7(3).

665 *Immigration and Asylum Act* 1999, s. 149(3).

666 *Sargeant v Macepark (Whittlebury)* [2004] 4 All ER 662.

667 LTA 1927, s. 25.

668 LTA 1927, s. 19(4).

669 LTA 1927, s. 19(4).

670 *Criminal Justice Act* 1991, s. 84.

671 *Criminal Justice and Public Order Act* 1994, s. 7(3).

672 *Immigration and Asylum Act* 1999, s. 149(3).

673 LTA 1927, s. 19(4).

674 *Sargeant v Macepark (Whittlebury)* [2004] 4 All ER 662.

675 LTA 1927, s. 25.

676 LTA 1927, s. 19(4).

677 LTA 1927, s. 19(4).

678 *Criminal Justice Act* 1991, s. 84.

679 *Criminal Justice and Public Order Act* 1994, s. 7(3).

680 *Immigration and Asylum Act* 1999, s. 149(3).

681 LTA 1927, s. 19(4).

682 The words in italics are to be repealed and replaced with the underlined words.

Index